EDITORIAL

Material for Future Issues

This issue, as you will have observed, has been produced under the benevolent eye of a new editor, Andrew Lambert having moved on to other fields of activity in which we wish him every success.

Although for the present we have a varied assortment of interesting articles on file, we are constantly on the lookout for fresh material of interest to naval enthusiasts. In this connection we feel sure that many of you who are reading this have something to offer. All we ask is that contributions are typed or carefully handwritten on one side of the paper with double spacing and a generous left hand margin. Although it may be years since you last put pen to paper, don't let that worry you: the editorial team are on hand to iron out any problems that may arise. We are also interested in any unusual photographs you may have acquired through service afloat or from relatives who fought in the two World Wars and the various minor conflicts that followed.

We would also like to see more queries in the editorial postbag; another journal in this field has a lively section devoted to 'Questions and Answers', and we feel *Warship* would benefit from an active feature of this nature.

To give you some idea of the material we are seeking, we are particularly interested in early marine engines and the engineers who attended them in the RN and foreign navies; the auxiliary warships taken up by the Axis Navies in World War II and their service histories; the captured units utilised by the Germans in Scandinavia and on the Russian Front; accounts of salvage operations on wrecks in ports

Can anyone confirm that this old torpedo boat is *Seestern*, ex-*Hvas*, of the Royal Norwegian Navy, seen here on German service sometime during World War II? *Courtesy L & L Van Ginderen Collection*

such as Narvik; anti-aircraft cruisers and anti-aircraft auxiliaries; minesweepers and minelayers; the final generation of China gunboats; HM cruisers *Effingham* and *Delhi* in their final form; the US Coastguard; the Confederate Navy; the 'Old' Indian Navy; The Focke-Wulf Fw 200 *Condor* and the German Naval Zeppelin Service of World War I; torpedoes and mines; ships versus aircraft; manning the navy; the mass production of naval ordnance during World War II and so on.

Finally, with regard to the series of articles on IJN *Tone*. The author has asked us to point out that there were a number of errors and omissions in the series. These occurred in the process of reducing the extensive and complex material to manageable form and were the responsibility of the editor not the author. We apologise to Herr Lengerer, his co-authors and our readers; standards of editing and proof-reading have been tightened up, so similar problems will not occur in future.

Ian Grant

The armour-clad at bay. Yesterday's threat countered by the quick firing gun, the search light and the torpedo boat destroyer. We would very much like to see an article on this theme. *CPL*

THE SALVAGE OF HM S/M *L55* BY THE SOVIET NAVY
The Reason Why

Przemysław Budzbon and Boris V Lemachko

View from forward after the hull had been cleaned.

Following the collapse of strong central government in Germany, Soviet Russia denounced the Treaty of Brest-Litovsk and launched a series of offensives against the small Baltic states and Poland. By the end of 1918 the Red Army had occupied large parts of Estonia and Latvia and was marching on Vilna, which was taken in the first week of the New Year.

The Estonians, however, assisted by a British Naval Squadron, repulsed the advancing Reds in the following three weeks, whilst the Polish Army was successful in containing them in the East in February and they were finally evicted from Latvia in March. The failure of the Red Army strengthened support for the White Russian faction which, backed by some western governments, launched offensives against Moscow from Siberia, the River Don and Estonia.

On 13 May 1919, the Northern Corps of the White Army under General Yudenich, supported by the Estonian Army, broke through the lines of the 7th Army of the Soviet at Narva and began an offensive drive on Petrograd. The initial impetus of the attack, backed by two naval operations, forced the 7th Army to retreat and within two weeks over 40 miles of territory had been captured. Faced by this counter-revolutionary threat to the birthplace of the Bolshevist Revolution, the Soviet for Workers and Peasants Defence, concentrated the few available reinforcements in the area and appointed Josef Stalin their Proxy Extraordinary to oversee the defence of the city. He was armed with special powers, the most sweeping of which was the right to discharge any civil or military officer for dereliction of duty and try them by a Court Martial.

It might well have seemed to the casual observer that the defenders held the trump card in the shape of the former Baltic Fleet – under its new name, Naval Forces of the Baltic Sea – which lay at Kronstadt under the command of the renegade Czarist officer Commander Zelenoj and consisted of four dreadnoughts, three pre-dreadnoughts, nine cruisers, 91 destroyers and torpedo boats, 22 submarines and over 140 lesser craft. The truth was that over two years of neglect had reduced most of the fleet to little more than scrap metal. Although it had been intended to man the most valuable units and form the Dejstvujushchij Otrjad (Active Squadron) known as DOT, the shortages of every sort in the dockyard had led to a considerable revision of

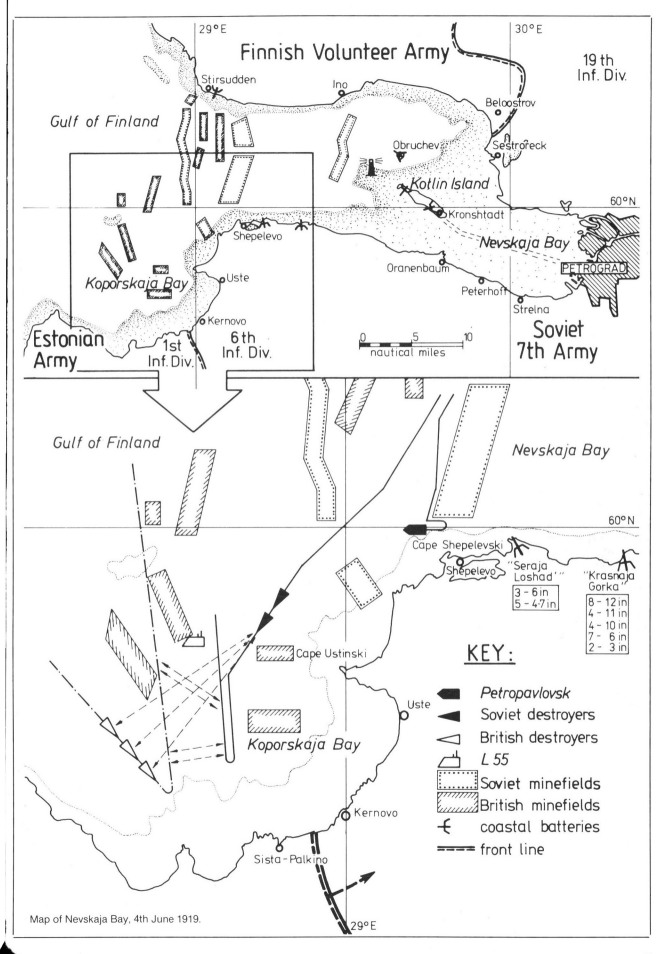

the planned establishment approved by the Revolutionary War Soviet on 23 January 1919.[1]

The effective strength available to Commander Dmitrev, commanding DOT at the end of May 1919, was no more than two destroyers, one submarine and 14 lesser craft. In view of the crisis, two dreadnoughts were fitted out and added to his command albeit with only half their boilers fit for steaming. Backed by three lines of defensive minefields totalling 2000 mines, and the fortress guns of Krasnaja Gorka and Seraja Loshad, this tiny force could only control the waters of Nevskaja Bay (the eastern section of the Gulf of Finland) which was sufficient to defend the roadstead of Petrograd. On the opposing side, the British Baltic Force of nine light cruisers, 18 destroyers, including two Estonian units, and ten submarines under Rear Admiral Cowan acted to prevent the DOT influencing the outcome of the land campaign. To contain the Red Fleet the British laid minefields across the southern entrance to the bay west of Cape Shepelevski, and under the guns of the Finnish battery at Stirsudden. Additional fields were laid in Koporskaja Bay (see map). Further out to sea the British maintained a constant watch with submarines and destroyers.

Enter the Royal Navy

At 1500hrs on the 4 June 1919, the lookouts at Krasnaja Gorka reported enemy destroyers and transports in Koporskaja Bay. Being concerned about the possibility of a landing behind the lines of the 7th Army, Commander Dmitrev ordered the First Destroyer Flotilla, Commander Rostovtsev, to sea on a scouting mission at 1545hrs. The destroyers *Gavriil*, Commander Sevost'janov, and *Azard*, Commander Nesvitskij, sailed immediately, followed by the dreadnought *Petropavlovsk* at 1600hrs. At 1645hrs, Commander Rostovtsev advised that the strength of the intruders was three destroyers only, with no transports. As the reports of a landing seemed to be unfounded, *Petropavlovsk* prudently stood by off Cape Shepelevski whilst the destroyers were ordered to engage the enemy (*Versatile*, *Vivacious* and *Walker*). The Soviet units steering southeast, *Gavriil* leading, then opened fire and subsequently changed course to the south to prevent the British bombarding the Red Army position on the coast at Kernovo. The British immediately turned north-northwest and the two Soviet craft, reversing course, did likewise at a distance of 25 cables (5000 yards approx), both sides exchanging fire throughout with no visible result. As the faster British destroyers drew ahead, the lookouts on *Gavriil* spotted two torpedo tracks heading for the ship. Commander Sevost'janov ordered full starboard rudder and full astern, which probably saved the day for the Soviet force.

The Destruction of L55

Subsequent events were reported by Commander Rostovtsev as follows: 'At 1737hrs two torpedo tracks were sighted to port, followed by the emergence of the bridge (conning tower) and part of a submarine hull at about five cables distance (1000 yards). The torpedoes passed ahead and I ordered the Commander to port his helm, intending to ram the submarine. Both destroyers turned towards the enemy and engaged with their forward guns. A huge column of black water erupted from the sea where the submarine had appeared, similar to that seen when a moored mine explodes, and fluttering debris was seen in the air. Heavy bubbling appeared on the surface, the escaping air generating white foam. The depth of water at the point where the explosion occurred is between 15 and 20 fathoms.' Following this success, the Soviet destroyers disengaged and returned to Kronstadt with *Petropavlovsk*.

Azard was credited with sinking the submarine as she had scored a hit on the conning tower and the subsequent explosion was attributed to this hit. Commander Rostovtsev's reference to the explosion of a moored mine was omitted by Commander Dmitrev in his report dated 5 June 1919. 'Our destroyers engaged the three enemy vessels at long range and sank an enemy submarine, which had attempted to torpedo *Gavriil*, with gunfire. A total of 82 shells was expended.' The sinking of *L55* by *Azard* has long been celebrated by the Soviet Navy as it was an early success of the newly formed Workers-Peasants Red Navy. The large oil spill in the area and the official announcement by the British Admiralty on 12 June 1919, confirmed the kill.

New Submarines for the Soviet Navy

From August 1923, the Soviet Navy was actively studying the question of submarine construction and the feasibility of such a project in the light of the available resources. The conclusion was reached that the Soviet Union did possess the necessary capability and the Soviet of Peoples' Commissars authorised work on the design of a modern submarine. In 1925, appreciable funds were allocated to the project and the design bureau, headed by Engineer Boris Malinin, was founded to bring the scheme to fruition. The major problem was a lack of practical experience as the last submarine designed and built in Russia dated from 1911 and was quite obsolete. Attempts to obtain foreign assistance were, not surprisingly, unsuccessful and progress was painfully slow when quite unexpectedly fortune smiled upon the design team.

In 1926, the gunsight from a British 4in naval gun was recovered during minesweeping operations in Koporskaja Bay. The news electrified Malinin as the wreck of *L55* was a veritable treasure trove of the latest submarine technology and would provide him with the answer to a problem that had long puzzled the Russians – the quoted surface speed of 17 knots. Thus, when on 26 November 1926, the Soviet of Work and Defence approved a five-year naval building programme, the decision was taken to raise and examine the wreck.

The first divers of the Ekspeditsija Podvodnykh Rabot Osobovo Naznachenija/Rescue Party for Underwater Works of Special Significance (EPRON) went down to the wreck on 20 October 1927, only a few days before the ice started to form in the area. Taking advantage of two days of relatively calm weather they carried out a preliminary survey of the wreck site. *L55* was found in position 59°55'10"N, 28°49'15"E. She was lying in 105 feet of water on a soft sandy clay bottom, buried up to her full width, inclined 7°/8° to starboard and trimmed aft by 4°. Except for a shot hole in the bridge (conning tower) no other damage was noticed due to the poor visibility and the lack of time available so late in the season.

The Salvage Operation

The hoisting weight of the wreck had been calculated at no more than 860 tons so it was practicable to use the submarine salvage vessel *Kommuna* with her maximum lift capacity of 1000 tons. The original plan was to lift *L55* in two stages; an initial lift to a depth of 65 feet for the journey to Kronstadt, and a final lift on arrival to pass her over the sill for docking. The salvage operation was scheduled to begin on the 15 May 1928. Four days later, divers from the Black Sea Section of EPRON arrived at Kronstadt and began a thorough inspection of the wreck. Only then was it discovered that the submarine had sustained serious damage to the bridge structure and the underlying pressure hull. The after part of the bridge together with the after 4in gun and parts of the casing lay 70 feet away to starboard. Under the port side of the wreck the sinker from a British pattern moored mine was found. As the strength of the pressure hull had been considerably reduced, all the calculations had to be revised which disrupted the timetable. Furthermore, *Kommuna* had been hired to EPRON for the 22 June only. It began to look as if the salvage team would not be able to complete their task inside the

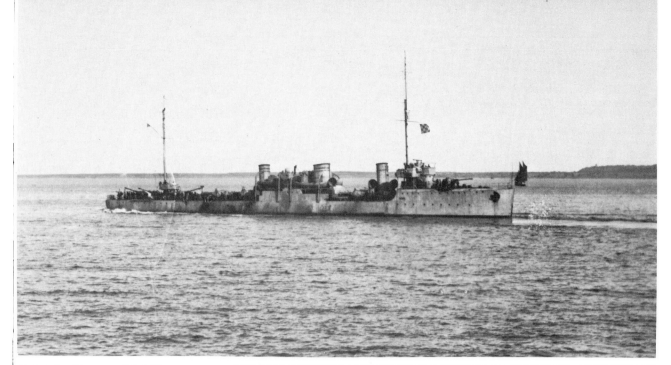

ABOVE: *Gavriil* as completed. She was mined with heavy loss of life in Koporskaja Bay on 21 October 1919. *Azard* avoided destruction on the same occasion, but was mined on 28 August 1941. (Renamed *Zinovev* 1922 and *Artem* 1934.)

RIGHT: Boarding the wreck.

short annual period of fine calm weather. These factors led to a revision of the salvage schedule and *Kommuna* was re-booked for a date in August.

Work began on the 18 June. The revised plan called for the insertion of a strong steel cradle under the midship section of the hull. The cradle consisted of four steel straps passed individually under the submarine. Fortunately, the nature of the bottom permitted the use of high pressure water jets for boring the necessary tunnels. The first was driven under the forward section and a steel cable passed through. The ends of the cable were taken aboard tugs on each side of the wreck, and by a similar series of operations three more cables had been passed under the hull by the 11 July. During the next ten days heavier cables and then the straps were hauled through. Prior to the tunnelling phase the tugs had worked a cable the full length of the submarine to break the bottom suction.*

As soon as the cradle had been assembled and secured to *Kommuna*'s heavy lifting gear on 10 August, preparations were made for the final lift. The weather was deteriorating and numbers of old mines were still present in the area. Four had made their appearance during the course of the salvage operations and been dealt with by the attendant mine-

* See notes

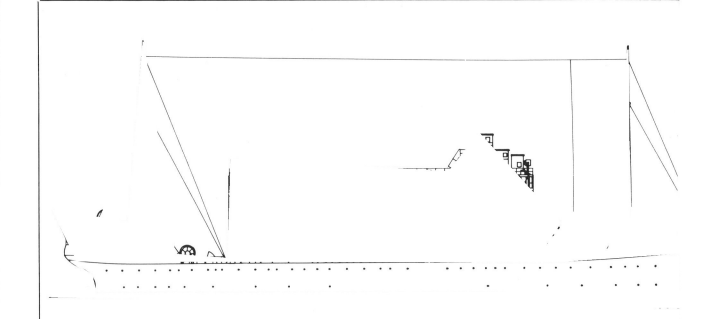

Volkhov as designed.

Close-up of damage, port side, note wooden salvage mat.

sweeper *Zmej*. In the circumstances it was decided that *L55* should be brought to the surface in one continuous lift, a somewhat risky proceeding if any of the gear failed. The order to commence lifting was given aboard *Kommuna* at 0623hrs on 11 August. Nothing happened until 0705hrs when the submarine began to move. At 0845hrs a mine broke surface on the starboard side of Kommuna and had to be gently pushed clear of the ship by a boat's crew. Finally, at 0945hrs the battered bridge of *L55* came into view after nine years on the bottom of the bay.

With the submarine slung between her twin hulls, *Kommuna* entered Kronstadt Naval Yard at 2300hrs. She docked the following morning and *L55* was lowered onto the prepared keelblocks, the cradle released and *Kommuna* hauled clear of the dock. The operation was completed by 1400hrs and by nightfall the dock was being pumped out. Next day the hatches were opened and the task of clearing the hull began. On 16 August the Commander in Chief, Naval Forces of the Republic, Commander R A Muklevich, announced the salvage of *L55* through TASS the official news agency and offered to return the bodies of the 38 crew to the British Government, an offer which was accepted via the Swedish Government, who sent a steamer to collect the coffins and deliver them to a British cruiser at Tallin.

Inspection of L55

The detailed technical examination of the submarine revealed that she had been sunk by a powerful external explosion. The port side of the pressure hull had been forced inwards for a distance of 28 feet immediately above the control room and extending five feet into the engine room. The engine room bulkhead was also bent aft, and the body of a member of the crew was found sandwiched between it and one of the plates of the pressure hull bent downwards by the force of the explosion. All the equipment in the control room was either shattered or displaced. In the engine room there was damage to the port engine cylinder head cover and pipework and fittings. Of the bridge structure (rather large in the *L50* Class very little remained. The forward 4i access trunk was exposed and the for part of the bridge, complete with gun. single periscope tube remained, but n trace of the conning tower and the aft 4in access trunk could be seen in th crumpled plating. The remains of th bridge casing confirmed the hit claime by *Azard*; an oval hole with burred edg was found there.

Cause of loss

In the light of the immediate eviden the Russians concluded that *L55* h been destroyed by a British moored mi Her commander had taken too grea risk in operating on the edge of a kno

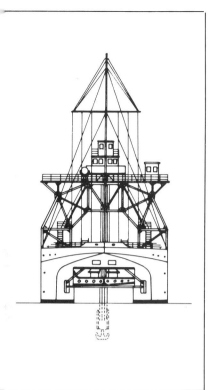

Damage as seen from starboard; Soviet officers inspecting.

minefield. After firing two torpedoes at the approaching destroyers, *L55* had inadvertently surfaced, probably due to a malfunction of the torpedo discharge compensation system. Knowing he had been sighted and aware of the hostile shellfire and possibly the hit on the superstructure, the submarine commander had dived into the minefield, catching the cable of a mine in his forward port hydroplane. The forward motion of the ship had pulled the mine down onto the port side of the bridge structure with devastating effect. The control room personnel had been killed instantly and as the bulkhead doors were open throughout the ship the remainder of the crew had been overwhelmed by the force of the explosion.

Survey and Reconstruction

On 13 August 1928, the Commander in Chief, Naval Forces of the Baltic Sea, Commander Viktorov, appointed Commander Voroblev, commanding officer of the submarine *Proletarij*, to *L55* in command. Although the first reaction of the Commission of Inspection was not very encouraging, the Soviet authorities were determined to reconstruct the submarine, regardless of the expense and the very real technical problems, including the lack of spare parts. At the end of August the submarine was towed to Leningrad and moored alongside a pier at the Ordzhonikidze/Baltic Yard, the builders of the *Dekabrist* Class submarines. The hull would appear to have been gutted for approximately 1000 working drawings were prepared and instruction manuals compiled, covering every aspect of the *L50* type submarine. The notebooks, belonging to her late officers were of assistance, but much of the material was obtained by trial and error.

Completion

L55 was completed on 7 August 1931 and for reasons of propaganda her name was retained, albeit in Latin style[2]. The first diving trials carried out in Kronstadt Middle Harbour alarmed the crew as she dived immediately and struck the bottom due to over enthusiastic use of the fast diving tank, a feature unfamiliar to the Russians. The speed trials proved especially disappointing as she only made 14 knots on the surface and 9 knots submerged instead of the 17/10½ reported in Jane's. (In fact, the Russians had done better than they realised for the *L50* Class had never achieved much over 14/8 knots. The figures quoted by Jane's were the optimistic design speeds. The fastest units of the type were those built by Armstrong for Yugoslavia and they did make over 15½ knots on trials. Ed.)

continued

Return to Service

She was officially commissioned as a unit of the Submarine Forces Brigade of the Naval Forces of the Baltic Sea on 5 October 1931. Nineteen days later she was lost with all hands during diving trials off Kronstadt due to the errors of an inexperienced crew. However, she was back in service by the end of the year with the newly formed Training Flotilla, to which she belonged for the next eight years. On 10 January 1940 she was reclassified as an experimental vessel. Her employment in this role ceased with the German invasion, and as she was not considered fit for active service she was laid up at Kronstadt at the end of June 1941. During the siege of Leningrad, the unreliable supply of electricity from the local power stations adversely affected the battery charging arrangements for the submarines of the Baltic Fleet, and on 11 May 1942, L55 was brought back into service as the mobile charging plant PZS2³. Her main engines and steering gear being in good condition, she proved very useful and was retained on strength after World War II. Her final years were spent in Tallin as L55 (Cyrillic form) and she was decommissioned and scrapped in about 1953.

Effect on Soviet Designs

A study of the design characteristics of the classes of submarine built in the Soviet Union subsequent to the detailed examination of L55 reveals certain features found in the British design with the exception of the Series I (*Dekabrist* Class) which were quite well advanced in 1928 and could only benefit to a limited extent, mainly in their internal arrangements, principally it is believed in connection with their ballast tank arrangements, the original layout not giving satisfaction.

The Series II (*Leninets* Class) marked a radical departure from the basic features of the previous class. The body lines were streamlined by reducing the midship section coefficient and decreasing the angle of entrance of the waterline planes. A straight stem was adopted, repeated in the Series II (*Shchuka* Class), and later classes, then abandoned in favour of the raked stem combined with a flared bow form to improve sea-keeping when surfaced. As the Russians anticipated, they would operate in confined waters where net barrages could be expected and the raked stem was better suited to cutting through nets than the straight form.

The double hull construction of the Series I type (960 tons) had given much trouble, during both the design and construction stages and it had also turned out heavier than expected. The Russians therefore were very happy to adopt the L50 form of hull with saddle tanks. The extra width improved surface stability as compared with the Series I and the hull weight was reduced by the introduction of light bulkheads. The narrow casing was adopted, which reduced top weight and improved diving times. This feature was only fully realised in the Series III and V designs however. Buoyancy tanks were added forward and deck ballast tanks abandoned.

Although the Russians were able to achieve savings in their hulls after the lessons learned by the careful examination had been digested, they were unable to save weight in other directions, the local version of British pattern fixtures and fittings turning out much heavier than the original. (For some reason the Russian finds it difficult to believe that equipment, if properly designed, can be both light and strong, an attitude of mind that inhibits the design process to this day.)

The British system of open battery tanks with wooden floors covered by rubber mats was adopted. In addition, a common system of battery ventilation was introduced and the use of hydrogen absorbent. However, Russian charging practice caused excessive hydrogen production and overheating of the batteries during the trials of *Leninets*, so the practice of ventilating each cell individually was reintroduced.

The Russians were particularly impressed by the motor/generators and the reducer arrangements which were copied and fitted to all their submarines. In conclusion, it can be said that L55 had a profound influence on the development of the Russian submarine. Her timely salvage enabled the Soviet Navy to bridge the gap in every sense of the word, and the Design Bureau exploited to the full the knowledge gained. By the early 'thirties, the Russian submarine represented the state of the art as at the end of World War I and further progress was blocked only through lack of access to the latest developments in engine design and foreign proposals in respect of the next generation of submarine craft. Both these obstacles were cleared in the first half of the decade. In 1933, the Soviet Government purchased the design of a medium size submarine from the Ingenieurs-kantoor voor Scheepsbouw, and the following year Hitler approved the export of MAN diesels to the Soviet Union. Armed with the latest in contemporary design and thinking, the Soviet Navy never looked back, and the further bonuses provided by the Allies during World War II enabled the Russians to take full advantage of the material seized in Germany in 1945. Nevertheless, the Soviet Navy still has its problems and these may well be related to the nature of Soviet society itself as Malinin discovered some 60 years ago.

RIGHT: Condition of hull after nine years immersion. Note shape of stern.

BELOW RIGHT: *Kommuna* (ex- *Volkhov*) post World War II. *Marek Iwardowski*

HM Submarine L55

A modifed *L* Class design to give improved torpedo armament forward and increased gun armament. Developed without any sea experience of the *L* Class; ordered some four months or so before *L1* was launched.

Builder: Fairfield Shipbuilding & Engineering Co Ltd, Govan, Glasgow. Laid down in early 1918. Launched 21:9:1918. Completed 19:12:18. Length: 235ft,0in (oa), 230ft,6in (pp), 206ft,6in (pressure hull). Beam: 15ft,8½in (pressure hull), 23ft,5½in (over saddle tanks). Depth: 15ft,9¼in (pressure hull), 16ft,8¾in (to underside of ballast keel). Surface displacement 960 tons. Submerged displacement 1150 tons. Saddle tanks 190 tons. Reserve of buoyancy 19.8%. GM (Design) 20in. BG (Design) 12in. Machinery: Twin screw, surface: 2400bhp. 380rpm. 14 knots (actual). 3780 miles at 12 knots (actual). Submerged: 1600bhp. 300rpm.8 knots (actual). 38 miles at 5 knots (actual). Armament: 6 bow tubes 21in. Torpedoes carried: 12. Guns: 2 × 4in. Periscopes: 2 × 30ft.

Main engines: Vertical heavy oil, four cycle, solid injection, designed by Messrs Vickers. Two sets at 1200bhp each. 12 cyl/ea 14½in × 15in stroke. Oil fuel consumption: ·43lb/bhp/hour. Total weight of main engines: 64 tons. Oil fuel stowage: 78 tons. Main motors: Twin armature, shunt winding. 800bhp per set. Voltage: 220. 300rpm. Rating: 1½ hours. Overload: 2000bhp for ½ hour. Motor weight per shaft: 25 tons. Batteries: 336 cells. Voltage in series: 220. In parallel: 110. Total weight: 135 tons approx.

Max diving depth: 250 feet. Weight of hull and fittings: 473 tons. Diving time at full buoyancy, both engines at cruising speed: 2min,10sec. Diving time at low buoyancy, both engines at cruising speed: 1min,45sec. Stopped in both states, the times were even better. No times recorded using the quick diving tanks; in fact, British sources do not refer to them. All the main ballast water was carried externally in 14 tanks in the saddle tanks and in three buoyancy tanks in the bow superstructure. The maker of the main engines is not recorded. Complement: 38 (Sources vary from 38 to 40/44). Although generally similar to the L Class, the stern

was of duck tail form and above water. The class proved awkward to surface due to cross-venting of the LP blows to the external tanks, later improved.

Kommuna - *Submarine Salvage Vessel*

Displacement: 2400 tons full load.
Dimensions: 315ft × 43ft,4in (oa), 11ft,10in (single hull) × 12ft (draught) *96 × 13.2, 3.6, × 3.7m.*
Machinery: Twin screw, two reversible diesel sets at 600bhp each = 10 knots. Two diesel generating sets at 80bhp each. Radius 4000nm/6 knots.
Complement: 250

Projected under the 1908 Estimates. Design based on the German *Vulkan*. Authorised in 1911. Laid down November 12th 1912 at the Putilov Yard, St Petersburg. Entered Navy List 9 July 1913, launched 17 November 1913. Twin hulls with common forecastle and poop. Four horseshoe form lattice girders linking mid-sections of twin hulls fitted with heavy duty electric winches 60 feet above load waterline. Total lifting capacity, 1000 tons. Maximum operational depth 130 feet, hoisting speed one foot per minute. Fully equipped with workshops; pumping capacity, 3000 tons per hour.

Commissioned in the Baltic Fleet as *Volkhov* on 27 July 1915. Renamed *Kommuna* 31 December 1922. Between 1917 and 1939 successfully raised the following craft: Submarines *Edinorog* and *AG15* (1917), *L55* (1928), *B9* (1933), *B3* (1935) and *M90* (1939). Patrol craft *Kopchik* (1923). Post-1939 recoveries not released; served as submarine depot ship 1940-45. Refitted and modernised after World War II and transferred to the Black Sea Fleet. In existence 1982.

NOTES:
1 Strength of DOT authorised by the Revolutionary War Soviet on 23:1:1919: Two dreadnoughts, one pre-dreadnought, two cruisers, ten destroyers, seven submarines, two minelayers, ten guardships and six minesweepers.
2 According to *Soviet Warships of World War II* by J Meister, *L55* was renamed *Bezbozhnik*. The authors have been unable to confirm this but think it possible between 1931-34.
3 Plavuchaja Zarjadovaja Stantsija/Floating Charging Plant.

REFERENCES (published sources)
1 A. V. Basov (ed), 'Boevoj Put' Sovetskovo Voenno-Morskovo Flota', Moskva 1974
2 V. I. Dmitriev, 'Atakujut Podvodniki', Moskva 1973
3 L. A. Emel'janov, 'Sovetskie Podvodnye Lodki v Velikoj Otechestvennoj Vojne', Moskva 1981
4 K. L. Grigajtus, 'Vosstanovlenie Podvodnoj Lodki L-55', Sudostroenie 2/1980
5 G. N. Kholostjakov, 'Vechnyj Ogon', Moskva 1976
6 A. M. Kosov (ed), 'Dvazhdy Krasnoznamennyj Baltijskij Flot', Moskva 1978
7 V. F. Kovryzhkin, 'Korabl' - Truzhennik i Voin', Sudostroenie 7/1970
8 A. M. Matijasevich, 'Po Morskim Dorogam', Leningrad 1978
9 G. I. Mocharuk, 'Sluzhba Podvodnykh Rabot Osobovo Naznachenija', Sudostroenie 11/1973
10 V. A. Molchanov, 'Vozrashchenie iz Glubin', Leningrad 1982
11 G. M. Trusov, 'Podvodnyje Lodki v Russkom i Sovetskom Flote', Leningrad 1963

*The authors have re-examined the sources and ascertained that only one tunnel was cut by waterjet, that at the bow; the first stage cables were pulled into position along the length of the hull. The steel straps took up a 'U' form under load.

The Story of the USS Vesuvius and the Dynamite Gun

Francis J Allen puts the record straight for this well known and misunderstood trials vessel

There was a period not so long ago when the US Navy was in such a state of disrepair that for all intents and purposes it had ceased being an instrument of national policy or defense. Salvation came with the advent of new thinking coupled with a fresh administration in Washington. That was all that was needed to start the Navy on the road to recovery. There was, of course, much work to be done, and not a few false starts. Due credit must be given to those men who followed the cautious approach in their rebuilding of the Navy, and formulated the strategic doctrine that was to bear fruit in the twentieth century. However, it is not the purpose of this article to dwell on the straight and narrow path but to look for a moment at just one of those false starts which added to the knowledge available for future generations.

In all candour, the ship which is the subject of this article was not greeted with any show of enthusiasm by the men who were initially entrusted with the task of directing the reconstruction of the Navy.

She was strictly an experimental vessel, her only purpose being to test, at sea, the experimental pneumatic gun, produced by the Pneumatic Dynamite Gun Company. The ship herself was built by the Cramp Shipbuilding Company of Philadelphia.

The Dynamite Gun

As for the gun, the underlying reason for an interest in a pneumatic weapon was what might be described as the embarrassing situation in which the navies of the world found themselves during the late eighteenth century. The embarrassment stemmed from the significant advances in the development of high explosives, which gave promise of improvement in the destructive power of naval guns but at the same time the explosives were found to be unstable. In a nutshell, the problem was that these new compounds tended to explode whilst still in the barrel of the gun due to the shock of

USS *Vesuvius* (1890-1922) Photographed during the 1890s. USS *Miantonomoh* (BM-5) is in the background. *US Naval Historical Center Photograph*

USS Vesuvius In 1891. *US Naval Historical Center Photograph*

A fine model of the *Vesuvius* held by the Smithsonian Institution. *Courtesy Smithsonian Institution*. All photographs courtesy of the Naval History Center.

firing. Later in the century a propellant was found which gave the shell a push instead of a kick. It was during the search for a solution to the problem that the idea of a pneumatic gun took hold. Inventive minds were hard at work and in the person of one Milford of Ohio the idea took on a practical form. Mr Milford was able to interest the government in his pneumatic device which used compressed air as the propellant in place of the commonly used black powder. A certain Lt E L G Zalinski was assigned to test and develop Milford's invention.

As the smooth bore gun barrel would consist of a very light tube and the firing pressure would not exceed 1000 pounds per square inch, the barrel, if made of steel or aluminium bronze, needed to be no more than one-half inch thick, even for large calibres. A balance valve was arranged so as to open and close by a single hand movement. This valve controlled the pressure in the gun itself. Air reservoirs were produced in the form of wrought iron, lap-welded tubes $12\frac{3}{4}$in to 16in in outside diameter and from $\frac{1}{2}$ to $\frac{3}{4}$in thick. These tubes were 18 to 20ft in length and were referred to as the firing reservoirs. The heavy duty air compressor for this system was manufactured by the Norwalk Iron Company.

The Charge

The torpedo shells were made of brass tubing and castings, which were as light as would be consistent with the necessary strength required for both handling and firing. The charge was made up of uncamphorated explosive gelatine, with a core of dynamite. It was found, during testing, that it was necessary to initiate detonation at the rear end of the charge, in order to produce the maximum effect on a solid target. Therefore, it was important that this should take place an instant before impact if the shell was travelling underwater. In the light of this requirement an electrical fuse was devised which consisted of the electric battery, the low tension primer, the circuit breaker, and the detonating cap.

The torpedo shell was designed for use both above and below water. Using wrought iron armour as the target for the uncamphorated explosive gelatine shell the following penetration values were obtained:

55 pounds	–	4.8 inches
100 pounds	–	6.6 inches
200 pounds	–	9.3 inches
400 pounds	–	13.2 inches
600 pounds	–	16.0 inches
700 pounds	–	17.4 inches
1000 pounds	–	20.4 inches

Firing Trials - Silliman Sunk

The question as to whether the pneumatic torpedo gun could deliver accurate fire or not would appear to have been settled by the trial before the Naval Board in June of 1886, and the destruction of the condemned schooner *Silliman* on 20 September, 1887. In the first case, four out of five shells hit roughly the same spot at a range of 1630 yards, and the fifth shell went over by about seven yards. With the *Silliman*, ten shells were fired in about ten minutes, 45 seconds. Had she been anchored end on, two of the shells would have struck the vessel, while four more would have exploded close enough to have destroyed her, and a further two would have seriously injured her.

Service Applications

A number of uses were envisaged for the pneumatic torpedo gun. One use was for the replacement of the torpedo tubes in torpedo boats with the pneumatic gun. Yet another use was as a weapon for use instead of the ship's ram, a tactical weapon which engaged the attention of many naval officers. An example that was widely quoted was that of the torpedo ram *Polyphemus*, a British warship of the time that was designed to close with an enemy and engage from 300 yards down to point of impact. The suggestion was made that a smaller vessel armed with a pneumatic torpedo gun would be capable of doing the same job, but with much less risk to both ship and crew, because she would operate at much longer ranges than the *Polyphemus*.

The dynamite gun cruiser, subject of this article, was first envisaged as a vessel mounting three guns of 15in calibre. They were to be placed abreast at a fixed elevation angle of 16°, with the muzzles projecting through the deck about 37 feet from the bow. The range of these weapons was to be at least one mile and each shell would contain 600 pounds of explosive gelatine, equivalent to 852 pounds of dynamite or 943 pounds of gun cotton. The ship was to make around 20 knots. The training of the gun was to be accomplished simply by turning the vessel. the rate of fire was to be two rounds per minute, with a full outfit of 30 large calibre shells.

Development of the pneumatic gun was to progress at a satisfactory pace under the direction of Lt Zalinski. The gun went through several models, all shore based, and these all showed that within their range limitations they compared well with rifled artillery.

Vesuvius *Receives her guns*

The guns designed and built for use aboard a warship were constructed of drawn brass tubing one-half inch thick and 55 feet in length with a diameter of 15in. Three guns were installed in the fore part of the trials vessel *Vesuvius* with their muzzles protruding from the deck to a height of 15 feet. They were set at a fixed angle of 18°. The gun tubes extended down through the ship to the projectile handling room and magazine on the lower deck. All the gun control valves were located in the conning tower. The initial air pressure multiplied by the duration of blast fed to the gun tubes effectively controlled the range of these weapons.

A feature of the design of the shells used in the pneumatic gun was that after hitting the water the shell would travel along the line of fire but at a depth of about ten feet. Fuzes designed to exploit this characteristic ran through three designs. The first was devised by Zalinski himself. This fuze used two separate battery powered electrical impulse firing circuits. The first of the circuits fired upon impact, while the second circuit was activated by salt water. These were the fuzes used with the guns during shore trials and were judged to be satisfactory.

The first sea trials of the pneumatic gun used a fuze which was mechanical in nature. It had to be set before being fitted to the shell. In the later sea trials a change was made to a fuze designed by a Russian named Rapieff. These fuzes incorporated all of the features of the Zalinski fuze but in addition would detonate on impact at any angle. They were very troublesome initially but after a period of development did perform satisfactorily.

The Projectiles

Three sizes of projectiles were designed by Rapieff. They were the full calibre shell which contained 500 pounds of explosive, a smaller sub-calibre round containing 200 pounds of explosive and finally a shell of only 50 pounds capacity. Whilst in flight the projectile was spin stabilised by the helical tail surfaces. Both the shell and its fins were isolated from the gun tube by means of wood or fibre sleeves which were placed on the body section and on alternate tail fins.

USS Vesuvius

The *Vesuvius*, the vessel built to house these new weapons, had the general lines of a fast pleasure yacht and an undesirably large turning circle. This problem was aggravated by a steering engine which was underpowered.

She had vertical triple expansion

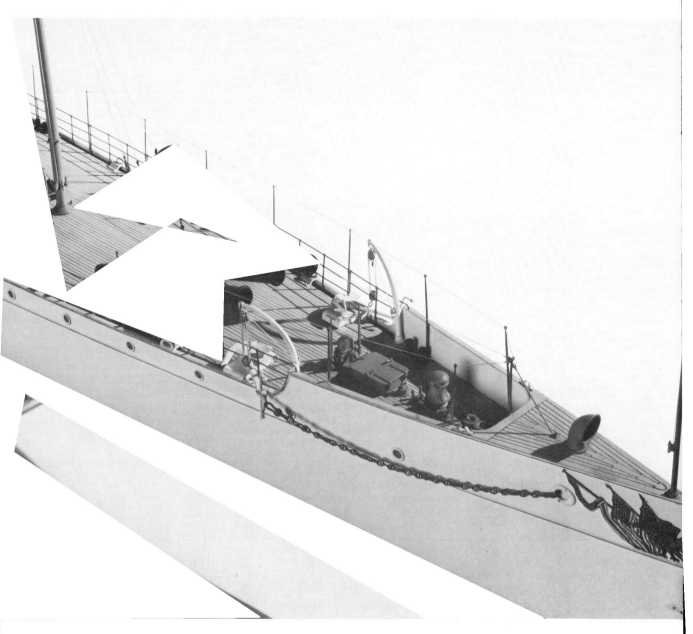

engines powering twin screws. Her boilers were four marine pattern locomotive type boilers which worked at a pressure of 160 pounds. In view of her design and, as it turned out, inadequate machinery, *Vesuvius* was simply not built to take full advantage of the potential offered by her armament. Her principal dimensions were:
   Length: 246 feet, 3 inches
   Beam: 26 feet, 6 inches
   Draft, mean: 9 feet, 3 inches
   Displacement at mean draft 805 tons

The *Vesuvius* was deemed ready for machinery trials in the winter of 1888. Her first trials revealed some minor problems but these were quickly fixed. The first set of official trials took place on 8 December 1888 but mechanical problems were at the heart of her poor showing. A second effort was made on the 27 December, but she still failed to make her designed speed due to problems with her machinery. Not until 21 January 1889, while negotiating a new course off the Delaware Breakwater, did *Vesuvius* attain a speed of 21.646 knots.

Further Gunnery Trials

The next order of business was her gunnery trials. These took nearly nine months as more shore based firings were thought necessary. As a consequence, *Vesuvius* was not ready for her final sea trials until 9 October 1889. On these trials she did well, well enough, in fact, to be perceived by the man in the street as the world's most powerful warship. The popular press even went so far to suggest that *Vesuvius* was just what was needed to deal with the Italian navy should Italy be so bold as to try to interfere in American politics. It was at about this time that *Vesuvius* went to sea for her gunnery trials. These were completed in due course and she was placed in commission in June of 1890. On commissioning she was listed as being armed with three pneumatic dynamite guns of 15in calibre, two 3in powder cannon, two 37mm revolving cannon and two Gatling guns. For two years *Vesuvius* was unemployed, but this state of affairs came to an end in 1891 when a second set of armament trials were conducted off Hampton Roads. The tests went well, and the most serious problem encountered was that firing had to be by word of command which gave the impression to those present that *Vesuvius's* guns were liable to a larger error in deflection than was the case. The shooting was seen as having been good but there was a general feeling of dissatisfaction with the boat as a whole. Even her skipper, Lt Seaton Schroeder, was concerned with the design of the vessel. He felt that *Vesuvius* could well have a place in the Navy as a vessel for shore bombardment but her design simply would not lend itself to that end.

Relegation to the Backwaters

She had been overtaken by improvements in the science of chemistry and technology and had become a design cul-de-sac. Unfortunately, the other less than satisfactory features of her design barred her employment in any alternative role. Quite simply, she was not worth spending money on, and spent the summer months showing the flag at the minor ports on the east coast where the public were suitably impressed by the mighty gun tubes and the seamen's yarns.

Her 'Finest Hour'

The Pneumatic Dynamite Gun Company went out of business and the number of projectiles in store began to dwindle. The outbreak of the Spanish American War led to a renewal of interest in *Vesuvius* and her 'fearsome' ordnance; accordingly, she was brought forward for service with the squadron blockading Santiago, Cuba. Following her arrival, she carried out blind firing after dark to the dismay of the Spanish defenders, and though she failed to hit anything the resulting fireworks were impressive.

Compressed Air Test Bed

At the conclusion of the war she was paid off whilst her future was under consideration. In 1905, she was commissioned as an experimental torpedo tender and served as such until 1922 when she was sold out of the service. On balance, she served her country well, reassuring nervous members of the public and fighting in the war, while no fatal accidents were reported in connection with the gun tubes and projectiles. This was followed by 17 years experimental work in compressed air technology.

Further reading: See Brassey, *The Naval Annual*, 1887 and 1888-89 and John D Alden, '*American Steel Navy*', 1972.

F7 lying in Oslo in May 1940. Note rubbing strake added and 3.7cm guns landed. Note trawler conversions on the opposite side of the quay.

F BOOTE OF THE KRIEGSMARINE
The Geleitboote Class of 1935

Mike Whitley

The thinking behind the design of one of the Kriegsmarine's 'bad bargains'

The ten vessels of this type have long puzzled students of naval history as they do not fit into the general pattern of warship classes. What was their intended role? Too fast for routine convoy duties and apparently ill-equipped for anti-submarine work, they remained an enigma. In the absence of positive information, foreign observers reluctantly classed them as convoy escorts.

In fact, they were intended as escorts for the three *Panzerschiffe* (see letter from Flottenkommando to OKM dated 3 November 1937). For this duty their speed can be seen in perspective, for the *Panzerschiffe* were only capable of 26 knots and the *F Boote*, being intended for the close escort, did not, in the eyes of the OKM, require any great excess of speed over their charges. Maintaining the outer screen, for which a good turn of speed was essential, was the task of the *Torpedoboote* and the *Zerstorer*. In addition, the OKM saw the *F Boote* as a fast minesweeper. For these roles a combination of good speed, shallow draught and reasonable radius of action was desirable. Anti-aircraft and anti-submarine duties were considered as secondary tasks, which made nonsense of the close escort concept. As designed, the class were experimental in several ways, particularly the hull form aft and boiler installation, both being features intended for use in the later *Type 34* destroyers. Displacement was set at 704 tons (Washington) but declared as 600 tons to avoid the charge of openly contravening article 8a of the London Naval Treaty. The external

F2, F4 and *F5* illustrating the two types of bridge fitted to the class.

appearance of the vessels was reminiscent of a destroyer without torpedo tubes. As a result, it was not surprising that they were sometimes mistaken for destroyers by British submarines. Internally, in view of their modest size, the layout was unusual in that the machinery spaces were divided into two self contained units with a consequent wide spacing of the funnels. Ideally, they should have been fitted with diesels for service compatibility with the *Panzerschiffe*, but to obtain the desired 28 knots the weight and bulk of compression ignition motors ruled them out, and steam turbine machinery was adopted. Unfortunately, the Kriegsmarine's passion for very sophisticated high pressure boilers proved to be the undoing of the class. Eight units were given La Mont boilers operating at 1616psi, and two (*F7* and *F8*) the slightly lower pressure boilers of the Velox pattern (1028psi). The two boilers, each in their own space, were separated by the forward turbine room and the auxiliary machinery space. Otherwise, the arrangements followed standard German practice with the forecastle given over to crew spaces, stores and magazines whilst the bridge structure housed the radio office, chart room, wheelhouse etc. The long deckhouses extending from the break of the forecastle to the quarterdeck provided useful accommodation, with the after end occupied by the wardroom. The officers' cabins were arranged aft of No 1 turbine room.

The main armament comprised two SK 10.5cm C/32 guns in single MPL C/32 mountings; one on the forecastle, the other on the after shelter deck. These pieces elevated to 50° and traversed 120° port and starboard forward and 130° port and starboard aft. Paired SK 3.7cm C/30 guns in LC/30 mountings were fitted on each side at shelter deck level just forward of the after funnel and two single 2cm MG C/30 were mounted in the bridge wings. Ranging and fire control were dependent on a single 3m rangefinder and a director sight on the bridge. Portable rangefinders were carried for the light flak weapons. The minesweeping winch was located on the quarterdeck as was the meagre depth charge outfit in single cradles.

Eight units were provided under the 1935 Programme and two the following year. Orders were placed for all ten of the class; six from Germania at Kiel, two from the Hamburg yard of Blohm & Voss and a final pair from Wilhelmshaven Navy Yard. *F1* joined the fleet at the end of 1935, followed by the rest of the group during 1936 and 1937. They formed the *1st and 2nd Geleiteboote Flotillen* based at Kiel under the command of *Führer der Minenschiffe* (FdM)/Flag Officer Mine-sweepers/Minelayers. *F1, 2,5,6,9* and *10* belonged to No 1 and *F3,4,7* and *8* to No 2. (The original intention was to build a total of 17 *F Boote*, but the final batch were replaced by 10 fast sweepers.)

Initial Shortcomings

By the Autumn of 1937, sea service had revealed certain shortcomings in the class which FdM formally reported to Flag Officer *Panzerschiffe* (FdP) in September. The report was critical of certain features of the design but concluded that seaworthiness was adequate. Not liked was the amount of spray thrown over the bridge in a seaway (alterations to the fore ends subsequently reduced this problem). Speed, not surprisingly, was found to be inadequate for the fast minesweeping role whilst acting as close escort except in a flat calm. Above sea state 4 they were unable to sweep effectively. Endurance was insufficient and the flak arrangements were poor even in good weather. On the positive side, FdM felt that they were acceptable as A/S vessels provided they had a sufficient margin of speed over their charges. In fact, he saw their main role as fast minesweepers for 14- to 18-knot merchant vessels provided their machinery did not let them down. F Kpt Rüge expressed the hope that in time familiarity with the boilers would lead to improved reliability and fewer breakdowns.

Failure of the Design

The following month Rüge submitted a further report, this time to *Befelshaber der Aufklarungsstreitkraft* (Flag officer Reconnaissance Forces) K Adm Boehm, in which he recommended the construction of an improved version as recent exercises had highlighted the need for fast sweepers. He foresaw a requirement for ten ships, eight to be based in the North Sea and the remaining pair to serve as Senior Officers' ships in the eastern and western commands. The existing *F Boote* could then be transferred to the less exposed Baltic with two allocated to the SVK or Boom Defence Command. At this time, another shortcoming of the existing design had come to notice, for despite the use of the redundant stabiliser spaces, fuel oil stowage remained barely sufficient for three days at 50/60 tonnes per day with a capacity of 174 tonnes.

In the light of the disappointing performance of the *Geleitboote*, Boehm felt obliged to redefine the role of the class to *Flottenkommando* (Fleet Command) and FdM. As before, they were to act either as fast escorts or minesweepers. The existing vessels were not successful because they lacked the necessary speed, endurance and seakeeping qualities. Nonetheless, until a suitable torpedo boat was available they still had their uses. He considered that new construction was an urgent priority and suggested that speed could be increased by reducing the armament, complement, superstructure and number of boats.

The basic criteria for the new design he listed as follows:

    Capable of 28 knots with sweeps out.
    Draught not greater than 2.5m.
    Two single 10.5cm with simple fire control.
    Three single 2cm/two single 2cm, one twin 3.7cm.
    Twelve units needed.

Fleet Command forwarded the report to the OKM (Naval High Command). The 1937 exercises had shown that the *F Boote* were not up to the task of screening the *Panzerschiffe*, but they were of use as minesweepers. No more of the class would be built and it was considered unlikely that an effective design could be produced within the (alleged) 600-ton limit. OKM concurred with an earlier suggestion that they be replaced by the older torpedo boats of the *Type 23* and *24* designs which would land the after gun, tubes and rangefinder to accommodate the minesweeping and A/S outfits.

Finally, in December 1937 a critical report on the class assessed them as follows:-

    Fast escort - too slow, endurance insufficient, sea keeping poor.
    Minesweeper - useful but draught too great.
    A/A capability - too lively for effective use as a gun platform.
    A/S capability - good but no detection gear fitted.

Partial Rebuilding of Some Units

As war approached, efforts were made to resolve the problems of the class, but new construction had priority in the shipyards of the Third Reich. *F3* was paid off for reconstruction on 28 October 1938, fol-

Bow view of an unidentified *F Boote* in striped camouflage.

lowed by *F6* next day. Both were taken in hand for conversion to Senior Officers' ships. The sides were plated from the break of the forecastle to the stern to give increased accommodation; the after 10.5cm and both 3.7cm twins were landed, the latter being replaced by two single 2cm. *F6* recommissioned on 20 September 1939 and *F3* on 5 March 1940. *F1*, *F2* and *F5* were paid off on 4 April 1939. Whilst *F1* became a Senior Officer's ship the other two were lengthened forward and equipped for torpedo recovery duties. *F5* recommissioned on 18 December 1939 and *F2* on 6 April 1940. Finally,

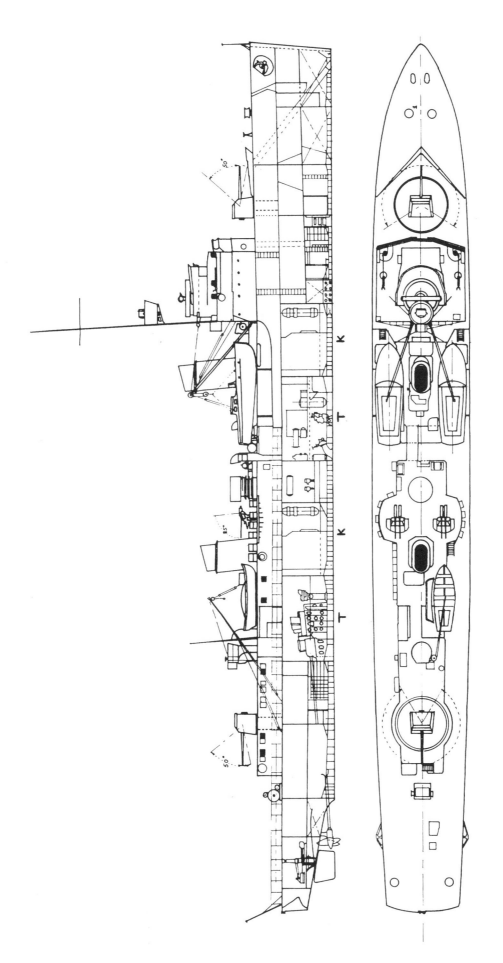

© M. J. Whitley, January 1987
F9 & 10

T = Turbine
K = Boiler

Three-quarter view of *F6* alongside with ship's company on parade for inspection by an admiral.

F4 paid off on 6 April 1939 and recommissioned on 7 August 1940, modified for experimental duties with torpedoes.

Service History

On the outbreak of war, only four units were in service (*F7-10*) with the *1st Geleitflotille*, F Kpt Pindter, in the Baltic where their initial task was patrol duty out of Swinemünde. During the first week of September, several suspected Polish submarine contacts were attacked without effect. The entry of Great Britain into the war forced the *Kriegsmarine* to divert its attention to the North Sea and on 6 September the flotilla passed through the Kiel Canal. Once in the German Bight, they screened the minelayers *Roland* and *Cobra* during the laying of the West Wall mine barrage. On several occasions suspected submarine contacts were attacked but with no better success than in the Baltic. *F9* experienced serious problems with her machinery and had to go into the Howaldt yard at Kiel for repairs. The other three units fared little better and because of the amount of time they spent in yard hands at Kiel, adjacent to the railway station, they soon became known as the '*Bahnhofs-Flotille*'.

F9 had only just rejoined when she and *F7* were ordered to sea to reinforce the escort accompanying the light cruiser *Leipzig* after she had been torpedoed by HM Submarine *Salmon* on the 13 December 1939. The two *F Boote* joined the escort on the morning of the 14th and took station on the port (*F7*) and starboard (*F9*) bow of the cruiser. Just off the mouth of the Elbe, HM Submarine *Ursula* attacked the group but missed *Leipzig*. One of her torpedo spread, however, struck *F9* in the vicinity of the bridge, blowing away her forepart. She listed to port and quickly went down by the head until her stern was vertical, and then sank. After she disappeared there was a series of explosions as her depth charges, magazines or boilers exploded. *F8* had to limp into Howaldt towards the end of the year and remained there, delayed by more urgent work and the harsh winter weather, until March 1940 when she returned to Cuxhaven. *F7* was similarly off station from January to at least March 1940. *F10* was transferred to Pillau in the east on the 9 January arriving on the 12th. Subsequently, she moved to Danzig. *F5* joined the flotilla early in 1940, but was under refit at Deutschwerke, Kiel, until March when *F8* joined her. As *F10* was immobilised in the ice at Gotenhafen, not one of the class was operational.

Relegation to Secondary Duties

War service had clearly demonstrated the short-comings of the design and FdM was determined to get rid of them. His report damned the class. In addition to the undesirable features reported earlier, he listed the lack of bilge keels, cracks in the shell plating, a tendency to roll wildly, defective bridge design and above all their constant mechanical breakdowns. If they were to be of any use, the bows would have to be rebuilt and bilge keels fitted, the hull stiffened with an external strake and the problems with the boilers and steam auxiliaries thoroughly sorted out. In connection with the boilers he suggested that the working pressure be reduced as had been done in *F1* and *F6*. On receipt of this catalogue of defects BSN recommended that the whole class be transferred to the Baltic for A/S duties and replaced by the 1st Minesweeper Flotilla. Whilst SKL agreed that they be replaced he did not concur with the proposed move of the *F Boote* to the east. Instead, the Maierform organisation was consulted regarding the problems of the unsatisfactory bow form fitted to the class. *F10*, scheduled for refit at the end of January 1940, was nominated as the trial hull. Four days later the *Marinekommandoamt* pointed out that the *1st Geleitflotille* was only four units strong (*F5, F7, F8,* and *F10*). Of the remainder of the class four were under conversion for other duties, leaving just *F4* available for minesweeping. As five units were too few to operate effectively, M Kdo Amt proposed that the flotilla be disbanded. SKL then reconsidered the whole matter and found that *F10* would have been in dockyard hands for over three months. There was a shortage of electric welders at Howaldt and unacceptable delays would occur in the completion of *Schiffe 45* (*Komet*) and the U-Boat programme. On 12 February the refits were cancelled and the ships allocated to other duties. The Flotilla was formally disbanded on 1 April 1940, and as FdM, Admiral Stohwasser remarked, 'it was a sorry story'.

War Service 1940-45

Following their removal from front line service the subsequent careers of the *F Boote* are not easy to trace. *F1* spent most of the war as a tender for Flag Officer Destroyers (FdZ) and after May 1942 spent some time in Norwegian waters when she moved north with *Lützow*, reaching Bogen Bay, Narvik, where she was briefly employed as an escort for the coastal convoys until her old problems beset her and she was ordered south to Trondheim on 26 August and then to Kiel where she arrived on 4 September. On the 15th she moved to Wilhelmshaven and remained there whilst the main destroyer and torpedo boat activity was centred on the French coast. In June 1943, she shifted to Wesermünde prior to leaving for the Baltic where she did the round of various ports from her detached base at Swinemünde. When Flag Officer Destroyers became responsible for the Skagerrak in 1944, *F1* proceeded to Aarhus and remained in Danish waters until surrendered at Copenhagen in May 1945. Although renamed *Libelle* and then *Jagd* in June 1944, she was referred to as *F1* in official papers until the end of 1944. Allocated to the USA in 1946, she served with the German Minesweeping Administration before being scrapped in France in 1947.

F2 spent the entire war attached to the 23rd and 25th U-Boat Flotillas at Danzig as a torpedo recovery vessel. Initially, she carried one 10.5cm gun but it was later removed. After the surrender she was allocated to the British and laid up in Scapa Flow, where she foundered at her mooring on 30 December 1946. The wreck was purchased for salvage by Metrec Engineering in 1967.

F3 served as a tender for FdM (Ost) in 1940, and FdM (Nord) by the summer of 1941. In addition, she served with the *14th Sicherungsflotille*. She subsequently moved to the Baltic and was refitted at Königsberg, returning to service as *Hai* and armed with a single 10.5cm UTO Flak L/45 on a U-Boat mounting. In her role as tender she served as flagship for Admiral Commanding Eastern Baltic at Libau. Withdrawn to the west in the face of the Soviet advance, she was sunk by rocket firing Typhoons of 2nd TAF off the Belts on 3 May 1945. *F4*, after a period with the SVK at Kiel, was serving with the 7th Minesweeper Flotilla in the North Sea by late 1944. Following the end of hostilities she was laid up at Lyness in Scapa Flow in 1946, and remained there until 12 January 1949 when she was handed over to the British Iron and Steel Company for scrapping.

F5, after a short detachment to Oslo in May 1940, served with the U-Boat Training Command following the disbandment of the *1st Geleitflotille*. Her armament was reduced to two twin 3.7cm and one 2cm. Eventually, she was returned to active duty and joined the 7th Minesweeper Flotilla. By January 1945 she was at Wilhelmshaven. On the evening of 29 January 1945 she activated an RAF magnetic parachute mine in *Weg 51* (the Swinemunde-Copenhagen swept passage). *F5* suffered heavy damage in compartments I–III but remained afloat and the buoy layer *Main* got a line aboard and attempted to tow her back to Swinemünde. Unfortunately though, she fouled a wreck, capsized and sank with 65 men killed or missing.

F6, renamed *Königen Luise* after conversion, was the only unit of the class to participate in the operation codenamed *Weserübung*, acting as flagship of Group 10, commanded by Flag Officer (Minesweepers) West, charged with the occupation of Esbjerg and Nordby on the west coast of Denmark. Later she saw service with the *4th Sicherungsdivision*, *4th Raumbooteflotille* and the *6th U-Jager Flotille* before being sunk at Wilhelmshaven on 30th April 1945 by B24s of the 8th USAAF.

F7 spent most of her time with the U-Boat Training Command, eventually joining the 7th Minesweeper Flotilla at the end of the war. She went to the USSR in 1946.

F8 spent a short period at Oslo in May 1940 after the *Geleiteflotille* was disbanded and then moved to Danzig for duty with the 23rd and 24th *U-Booteflotillen*. She saw some brief action on A/S duty when she joined the hunt for the Soviet submarine *L21* after the latter had met with some success off the Stolpe Bank. Ceded to the USA after the war, *F8* saw no further employment and was broken up at Arnbacht by Hendrik Ido in 1950.

F10 also attached to Flag Officer (U-Boats) was serving on escort duties in the eastern Baltic by 1944. In June of that year she formed part of the screen for *Prinz Eugen* during the sortie to the Aaland Islands. Found non-operational at Frederikshaven as a unit of the 7th Minesweeper Flotilla in January 1945, she was allocated to the USA and scrapped with *F8* at Arnbacht.

Postscript

An improved class of Geleitboote was designed and laid down but none entered service. With displacement increased to 1739 tons full load and a longer, wider hull the new class promised to be much superior to the *F Boote*, especially as they were given a well proven triple expansion machinery layout. The main armament was to have been two twin 10.5cm mountings, backed up by a good flak outfit. In addition, they were to have carried a helicopter* – a world first. Twenty-four units were projected – *G1* to *24*. The first four were ordered from Stulken at Hamburg as their yard numbers 789-792 on 14

---

* Probably a Flettner Fl 282 Kolibre (Hummingbird). The Kriegsmarine held a series of trials with these machines in the Baltic in 1942, and later used a small number on convoy protection in the Mediterranean and Aegean in 1943. An order for 1000 machines was placed with the BMW factory in 1944, but production was frustrated by the Allied bombing offensive. Can anyone throw light in this dark corner?

Table 1 Class Technical Details	
Displacement	712 tons (Std), 1028 tons (Full load)
Length	239½ft pp, (73.5m); 249½ft oa, (75.94m)
Beam	28¾ft (8.8m)
Draught	8¼ft (2.59m)
Machinery	Two La Mont boilers, except *F7* & *F8*, Velox type. Two shaft Brown-Boverie geared turbines, 14000shp=28kts 216 tons oil fuel=1500nm at 20kts
Armament	2, 10.5cm SK C/30; 4, 3.7cm SK C/30; 2,2cm MG C/30.
Complement	121

Note: After lengthening, *F1* and *F6* displaced 768/1065 tons, with an overall length of 263·12ft (80·20m).
Armament varied as described in the text and a few received twin 2cm LM44 mountings in 1944/45.
As built, the class were fitted with a 'Frahm' stabilising system later removed. It is believed to have been of the anti-roll, fluid transfer type. With regard to *F5*, the author is fairly certain she was lengthened although he has been unable to trace a post 1940 photograph of her.

Detail view of *F8* surrendered at Copenhagen at the end of the war, note that she still retains the twin 3.7cm amidships, and the assortment of surrendered craft in the photograph.

October 1941. Orders for the majority had been placed with Dutch yards on the 11th. *G5* to *8* went to Wilton Fijenord, the first two being yard nos 702–703. *G9* to *12* went to P Smit, Rotterdam; *G13* to *13* to De Scheldt, Vlissingen; *G16* to *18* to Werft Gusto, Schiedam; *G19* and *G20* to J & K Smith (Kinderdyk); *G21* and *G22* to Nederlandsche Dok, My; *G23* and *G24* to Gbdr Boele, Bolnes.

Work began on the first six only with *G1* planned for completion by 1 November 1943 (as at December 1941). Due to shortages of material and work with a higher priority, progress was slow and in April 1942 it was suspended for two months. Construction then resumed but finally ceased in May 1943. *G1* was wrecked on the stocks during a raid by the RAF on 27 July 1943, whilst the other five were dismantled later. None of the class entered the water.

Table 2 SHIP HISTORIES

Name	Laid Down	Launched	Completed	Fate
F1	2.8.34	1.3.35	15.12.35	To USA, 1946. Broken up in France, 1947.
F2	7.8.34	2.4.35	27.2.36	To UK, 1946. Foundered in gutter Sound, Scapa Flow, 30.12.46.
F3	22.8.34	1.6.35	7.3.36	Sunk by air attack north of Kiel, 3.5.45. Raised & b/u 1948.
F4	22.8.34	2.7.35	5.4.36	To UK, 1946. Handed over to BISCO for scrapping 12.1.49.
F5	6.9.34	14.8.35	1.5.36	Mined and sunk off Swinemünde, 29.1.45.
F6	6.9.34	1.10.35	25.5.36	Sunk by air attack at Wilhelmshaven, 30.3.45.
F7		25.5.35	15.2.37	To USSR, 1946.
F8		27.7.35	8.4.37	To USA 1946. Scrapped by Hendrik Ido, Arnbacht 1950.
F9		11.5.35	10.2.37	Torpedoed by HM/SM *Ursula* off Heligoland 14.12.39.
F10		11.5.35	12.3.38	To USA 1946. Scrapped by Hendrik Ido, Arnbacht 1950.

HERMES AT WAR

The Career of *Vasilevs Georgios I* 1939-43

Pierre Hervieux

The story of the only British-built modern destroyer to serve under the German flag during World War II

In February 1937, two destroyers were laid down for the Greek Navy by Yarrow & Co at their Scotstoun, Glasgow yard. They were named *Vasilevs Georgios I* and *Vasilissa Olga*. The former was launched on 3 March 1938, and the latter on 2 June of the same year. Both were delivered in February 1939. They were armed with four 5in, four 37mm and eight 21in torpedo tubes. In appearance, they were very similar to the *I* Class of the RN. Following the outbreak of World War II in September 1939 and the growing hostility of Fascist Italy, both were actively employed in Greek territorial waters. When Italy entered the war on the side of Germany in June 1940, the Greek Government instituted a Neutrality Patrol. Following the sinking of *Helle* by torpedoes of Italian origin off Tinos on the morning of 15 August tension grew between the two governments. There had been previous attacks by Italian aircraft on units of the Greek Navy and the two new destroyers despatched to Tinos in reaction to the attack on *Helle* were subjected to a high level bombing attack by a single aircraft off Syros. As a result of these incidents Greece quietly mobilised her armed forces and the Italians met with unexpectedly determined resistance when they invaded across the Albanian border on 28 October 1940. As is well known, the war went badly for the Italian armed forces, particularly the Army, ill led, poorly trained and not equipped to fight other than a colonial war. Inevitably, the Germans were drawn into the conflict although they were not at war with Greece. On 23 March 1941, German aircraft attacked an Anglo-Greek convoy on passage from Piraeus to Alexandria. Two of the five empty merchantmen were sunk, but the escort commanded by Captain Mezeviris in *Vasilevs Georgios I* was unharmed and the surviving ships made Alexandria safely.

Hermes was painted plain light grey between her commissioning in March 1942 and the end of that year.

Hermes in Suda Bay, Crete, with the wreck of the British heavy cruiser HMS *York* in the left background.

The German Invasion

Germany invaded Greece on 6 April 1941 and the Greek Navy was attacked wherever the Luftwaffe could find it. On the night of the 13th, *Vasilevs Georgios I* was steaming slowly, close inshore, near the village of Sofiko, Peloponnesus, where the high sea cliff afforded some protection. The night was very light and enemy aircraft were active in the area. The destroyer was spotted and attacked by a single aircraft which dropped a stick of bombs, one of which just missed her, causing heavy flooding and other damage. Listing at an angle of 30°, she crawled into Salamis at dawn and was placed in a floating dock for emergency repairs. During an air raid on the 20th, the dock was damaged by a bomb and efforts to move it to deep water for scuttling failed. Both ship and dock were damaged by demolition charges but the Germans had recovered the destroyer by the end of May and refitted her over the next nine months. *Vasilissa Olga* left Athens for Crete with the Greek Government on 22 April 1941 and subsequently reached Alexandria.

Under the German Flag

Repaired at Piraeus, *Vasilevs Georgios I* was given five 20mm mounts to boost her flak outfit and commissioned in the Kriegsmarine as *ZG3* under the command of Kapitän zur See Johannesson on 21 March 1942. Thus she became the largest German warship in the Mediterranean, having a displacement of 1414 tons and a complement of 225. As she had been designed to carry 127mm guns of German manufacture there were no ammunition supply problems. During the second phase of her career, *ZG3* served her new masters well. She was first employed in the Aegean from June 1942 until the beginning of April 1943. On the 22 August 1942 she was renamed *Hermes*, and it seems the name brought her luck. In the main she was employed on convoy duty between Piraeus and Crete and the Dardanelles with an occasional trip to Tobruk. As an escort leader, *Hermes* was highly regarded, always in the van and ready to assist damaged and sinking ships despite the frequent air and submarine attacks.

On the 2/3 July 1942, she escorted the German minelayer *Bulgaria* on a lay in the Aegean. The following month she went to the assistance of *U83*, Kapitänleutnant Kraus, which had sunk the Canadian transport *Princess Marguerite* (1925, 5875 tons) off Port Said on the 17th. Later the same day, the U Boat was caught on the surface off Haifa by British aircraft and damaged. Unable to proceed under her own power she requested assistance. *Hermes* rendezvoued with her next morning and towed her to Salamis. A difficult escort was completed successfully between the 22 and 24 September when *Hermes* and the Italian TB *Orsa* escorted the tanker *Rondine* from Suda Bay to Tobruk. Twice whilst on passage from Crete to Cyrenaica the tanker lost most of her power. Although attacked repeatedly by allied aircraft, the three vessels arrived unscathed in Tobruk and the tanker unloaded her cargo of motor fuel for the Afrika Korps. Between the 9 and 11 October, *Hermes* carried out a minelaying mission south of Crete. At the beginning of November, in company with the Italian destroyers *Freccia* and *Folgore*, torpedo boats *Ardito*, *Lupo* and *Uragano* she escorted the tanker *Portofino* and freighters *Mualdi* and *Col di Lana* from Piraeus to Benghasi. Although subjected to air attack, no losses were sustained on passage. On the 16th, whilst escorting the tanker *Celeno* and the freighter *Alba Julia* from Piraeus to the Dardanelles, she located a submarine at 1610hrs, north of the entrance to the Di Doro Channel. Also present were the submarine chasers *UJ2101* and *UJ2102*. Whilst the convoy steamed on, *UJ2102*, Oberleutnant Kleiner, followed up the contact, dropping a total of 49 depth charges during the hunt. Torpedoes were observed at 1942 and 2158hrs; at 2200hrs the submarine broke surface and fired at the submarine chaser. *UJ2102* returned fire and rammed her at 2205hrs and again at 2212hrs. The submarine, damaged both by the depth charging and the ramming, sank at 2235hrs. The Germans picked up 33 survivors including the Captain who revealed that his boat was the Greek *Triton* (1928, 760 tons) with 56 men on board including a British liaison officer.

Into the Fray

By the end of January 1943, the British Eighth Army was advancing on Tripoli and the Axis forces were retreating along the coast. At this point the German OKM decided to switch *Hermes* to the protection of convoys carrying supplies and reinforcements from Sicily to Tunisia. It was dangerous work due to the allied superiority in the air and the concentration of naval forces in the area - the Italians called it 'La rotta della morte' (The speedway of death!). At Piraeus, on the 2 April 1943, Kpt zer See Johannesson handed over command of *Hermes* to Freg Kpt Rechel, former Captain of *Z29* (9 July 1941-31 March 1943). She sailed immediately, passing through the Corinth Canal, the Ionian Sea and rounded the toe of Italy to reach Salerno on 4 April. On the 16th she experienced her first air raid since leaving Piraeus. A further raid took place the following day as *Hermes* loaded mines in preparation for a lay planned for the night of the 19/20 April in the middle of the Sicilian Channel. She sailed at 1530hrs on the 19th escorted by Italian MTBs. An air raid warning was in force but the aircraft proved to be a number of the clumsy six-engined *Gigant* Me 323 transports. Around midnight, no longer under the lee of the coast of Sicily, the combination of wind and sea was causing her to roll from 15 to 20° with her full deckload of mines. Half an hour later 'Deutsches Marinekommando Italien' warned her that her presence had been detected. A further message at 0116hrs advised that airborne reconnaissance had spotted three enemy destroyers on a converging course. Reacting to this report, *Hermes* altered course and at 0142hrs Rechel was warned that an American source was reporting an enemy destroyer approaching Malta. The lay commenced, as ordered, at 0335hrs and was completed by 0415hrs. *Hermes* then made for home. At first light a British Wellington bomber was sighted about a mile distant, but the ship was not seen. At 1000hrs *Hermes* worked up to full speed as Rechel was anxious to check her performance. She produced a shaft speed of 300rpm corresponding to a sea speed of 30 knots, and at 1416hrs was back in Salerno.

Sinks HM S/M Splendid

On 21 April 1943, whilst on passage from Salerno to Pozzuoli, a lookout sighted a periscope about a kilometre away to starboard and slightly ahead of the ship. *Hermes* closed with the sighting and obtained a positive contact almost immediately. Between the time of the initial sighting at 0838hrs and 0924hrs Hermes made three runs, dropping a total of 54 depth charges. The submarine then surfaced and *Hermes* engaged with main and anti-aircraft armament, scoring a number of hits and forcing the crew to abandon ship. The submarine sank stern first at 0947hrs, 20 survivors being picked up by *Hermes* and another ten by the Italian submarine chaser *VAS226* which came out from Capri. The submarine commander, Lt I L M McGeoch, was picked up by *Hermes* and identified her as HM S/M *Splendid* which had sailed from Malta on 18 April with a complement of five officers and 40 ratings. By 1200hrs, *Hermes* was alongside in Pozzuloi. An hour later Admiral Pini arrived from Marina Napoli to decorate Freg Kpt Rechel with the Italian Silver Medal for Bravery and then, following the ceremony, *Hermes* returned to Salerno. On the 23rd Rechel and his crew received a personal signal from Field Marshal Kesselring congratulating them on their success. They were busy loading mines for another lay codenamed Munition and were scheduled to sail at 2355hrs. At the last minute the operation was cancelled and a trooping run to Tunisia substituted. The full mine outfit had to be unloaded. These new orders arrived by telegram from Marina Napoli on the 25th, instructing '*Hermes* to transport 350 troops from Salerno to Tunis. To Gaeta at 20 knots in company with Italian destroyers *Pigafetta* and *Pancaldo*. Then to Trapani, from there towards Pantelleria, Cape Ras Mustafa and Cape Bon at 22kts. Must be in Tunis by 1200hrs April 26th. After unloading troops, return to Zambretta. Proceed to Pozzuoli for refuelling. Report arrival, switch lantern on'. Hermes sailed at 1710hrs, and at 0120hrs a small bomb exploded to port, some three minutes after the moon had made its appearance and aero engines were heard. Other aircraft were both seen and heard in the distance until 0440hrs when Supermarina signalled an increase in speed to 24 knots. A few minutes after sunrise four friendly fighters appeared overhead to provide air cover. The African coast appeared at 1138hrs and at 1158hrs speed was increased to 28 knots. The three destroyers arrived in the Roads of La Goulette at 1345hrs. After the troops disembarked, the wounded were ferried to *Hermes* by a sailing vessel and two Siebel ferries. The three destroyers sailed at 1542hrs, and aboard *Hermes* were 173 casualties and medical staff including 100 slightly wounded. At 1707hrs, the first hostile aircraft were sighted, but the three vessels were concealed by a fog bank; but not for long for at 1752hrs they were attacked by at least ten fighter bombers. *Hermes* brought her full flak outfit into action, whilst the Italians fired everything they had including their HA/LA main armament. None of the group suffered either damage or casualties, although one bomb dropped 50 metres away from *Hermes* during the five-minute raid. By the morning of the 27th they were safely at anchor off Trapani, concealed within another fog bank. *Hermes* arrived at Pozzuoli on the 28th and disembarked the party of wounded; she then refuelled and sailed for Salerno. That evening, Rechel reported a boiler defect and requested permission to carry out repairs, estimating they would take until the 30th.

The Final Sortie

Next morning, Deutsches Marinekommando Italien ordered *Hermes* to sea for a priority mission, the boiler defect to be made good on her return. Rechel advised his senior engineer accordingly and ordered the ship to be made ready for sea. During the afternoon of the 29th, *Hermes* embarked 213 troops of the Herman Göring Brigade and sailed for Pozzuoli at 1845hrs where she joined the Italian destroyer *Pancaldo* with more troops embarked and under her command sailed for Tunis. Nothing untoward occurred until 1012hrs the following morning when they were set upon by about 37 fighter bombers. A few men were wounded aboard *Hermes* and the after mast damaged during the ensuing action. At 1125hrs, a second attack developed when 18 twin-engined bombers with about 15 fighters as escort dropped bombs from about 200 metres without result. At 1136hrs, 12 twin-engined bombers made a further attack and scored a hit on *Pancaldo* some two to three miles off Cape Bon; she caught fire and came to a stop. Many of the troops aboard her took to the water. At 1210hrs, *Hermes* was attacked by about 18 fighter bombers but was not hit. Fourteen minutes later she was subjected to a determined attack by a further wave of 30 fighter bombers and several near misses caused problems with the starboard turbine and the helm which now failed to answer. At 1250hrs, 16 fighter bombers swept in for the kill. There were further near misses despite the volume of flak put up by *Hermes* and the port turbine lost power. Finally, at 1315hrs both turbines failed and the ship came to a stop. Some 60 troops were disembarked by using the rafts on board and they made for the shore. The Italian hospital ship *Aquileia* appeared at 1500hrs, picked up the occupants of the rafts and took on board seven wounded ratings from *Hermes*. Three motor sailing vessels arrived at 1555hrs; one took the crippled destroyer in tow and the four craft crawled towards the African coast. Twenty-five minutes later the small tug *Carthago* and a harbour protection launch joined the company; the tug took over the tow and the motley collection reached the coast near Corbus without interference at 1755hrs. Anchored close inshore, *Hermes* disembarked the remaining 160 troops with their equipment and the motor sailing vessels landed them in La Goulette. The weather deteriorated on 1 May and *Hermes* was forced to remain at anchor. She was attacked by 15 fighter bombers but without affect. On the 2nd she was attacked three times between 1110hrs and dusk. At 2030hrs the 600hp *Carthago* got under way at three knots and the two vessels reached the eastern roadstead of La Goulette at 0130hrs on the 3rd. By 0900hrs she was in Tunis harbour.

The End of the Road

Of the three destroyers that had sailed from Pozzuoli she was the sole survivor; *Pancaldo* had succumbed to direct hits, and *Lampo* coming up astern with ammunition had also been sunk. *Hermes* had survived due to a combination of luck, fine ship handling and an excellent flak outfit. She claimed no less than eight fighters and one bomber shot down during the fighting. A thorough examination of the ship soon revealed that she required skilled dockyard attention if she was to sail again, and Rechel was ordered to scuttle her to maximum effect in Tunis harbour. On 7 May she was sunk by demolition charges at 0832hrs across the harbour mouth. Initially, she settled on an even keel, but at 0910hrs she heeled to starboard and capsized 15 minutes later, lying half a metre below the surface. Rechel sent one final signal to Deutsches Marinekommando Italien and the OKM. 'Sinking of destroyer *Hermes* carried out. My thoughts go to my valiant crew, particularly to the 120 men I must leave behind in Tunisia'.

Thus ended the unique story of a British-built Greek destroyer in German hands. She was the only German destroyer to be given an official name during World War II, the only one to serve in the Mediterranean, and the only German-manned destroyer to sink two allied submarines. Her sister *Vasilissa Olga* did not long survive her. About four and half months later she sailed from Alexandria in company with HMS *Intrepid* for Leros in the Dodecanese. They arrived at 0500hrs on 26 September 1943 and were attacked by Ju88 bombers of the German Air Force two hours later. *Vasilissa Olga* was sunk in a matter of minutes, whilst *Intrepid*, seriously damaged, succumbed to another air raid later in the day.

Miscellaneous Notes

After the war it was revealed that *Helle* was sunk by the Italian submarine *Delfino*, Lt Aicardi, at the instigation of the Italian Governor of Rhodes and the Aegean Islands, De Vecchi, an impulsive personality. The sinking did not meet with approval in Rome and Aicardi was quietly removed from his command.

The additional 20mm mounts fitted to *Hermes* appear to have been singles, one on the forecastle, one each side of the bridge on a level with 'B' gun and one each side of the main or after mast on a level with 'X' gun.

The Me 323 *Gigant* transport was developed from the very large glider of the same name and proved to be a difficult beast to fly. It could carry 130 combat troops on two decks (more in an emergency) or 21,500lbs of freight. They were heavily armed but vulnerable due to low speed and poor handling characteristics. They flew from Trapani, Sicily, to North Africa, and some 20 of them, loaded with petrol, were shot down by allied fighters on the 22 April 1943.

HM S/M *Splendid* sank *Aviere* on 17 December 1942. Lt McGeoch mistook *Hermes* for one of the ex-Yugoslav destroyers in Italian hands. The submarine manned her gun on surfacing but the German fire proved murderous. Further information would be welcomed on the auxiliary naval and merchant vessels mentioned in the article.

Additional information provided by the author:

Submarine Chaser UJ2101, ex-*Strimon*, ex-*Cape Otway* (trawler, 325grt, 1925), Kpt Lt Vollheim, sank the Greek submarine *Katsonis* in the Aegean shortly after 2000hrs on 14 September 1943. The submarine appears to have been on the surface for she was heavily damaged by gunfire and then rammed. The action took place at the north end of the Euboea Channel near the small harbour of Trikeri and UJ2101 picked up 14 survivors including Commander Laskos and his RN liaison team – one officer and a wireless operator. UJ2101 was bombed and sunk by allied aircraft whilst on convoy duty, north of Crete, on 1 June 1944.

Hermes seen from a U-Boat, possibly U83, after she had towed the damaged submarine to Salamis in August 1942.

UJ2102, ex-*Birgitta* (trawler), Kpt Lt Kleiner, was sunk in action with HM Ships *Termagant* and *Tuscan* southwest of Cassandra-Huk on October 7 1944.

Bulgaria (1894, 1108 tons) was the former Bulgarian steamer of the same name. She is believed to have carried the following flak outfit: two 88mm plus several 37mm and 20mm. *Bulgaria* laid several minefields off the west coast of Greece and in the Aegean between May and August 1943. Whilst on passage to Cos, loaded with stores, she was torpedoed and sunk by HM S/M *Unruly*, Lt Fyfe, south of Amorgos, on 8 October 1943.

The *Celeno* was unsuccessfully attacked by the Soviet submarine *D4* between Constanza and Sevastopol on 1 June 1943. *Alba Julia* was a Rumanian vessel of 5700 tons, employed as a troop transport between Sevastopol and Constanza during the siege in April 1944 until her destruction by fire following bombing by Soviet aircraft on the 18th. *Rondine* (1924, 6500 tons) was a motor tanker. *Col Di Lana* was a single screw motor ship built by Cant Nav Triestino in 1926. 5900 gross tons, 417 × 53 × 25ft. Draught 24½ft. 11/13 knots.

The two ships sunk by JU88 bombers, whilst on passage between Piraeus and Alexandria with Convoy AS21, 20 miles southeast of Gaudo on 23 March 1941, were the Norwegian tanker *Solheim* (1934, 8100 tons) and the Greek freighter *Embiricos Nicolaos*.

AIR POWER
The Sinking of IJN Battleship Musashi

Tim Thornton

> The sinking of the Japanese super battleship and the vital lessons learnt by the US Navy

At 1935hrs on 24 October 1944 in the Sibuyan Sea, surrounded by the lofty volcanic islands which make up the Philippine archipelago, the giant battleship *Musashi* gave a sharp lurch to port and then rolled completely over floating bottom up. She had no sooner steadied in this new attitude than she began to slip smoothly beneath the surface bow first. Of her complement of 2279, some 952 went down with her including 38 officers and her Captain, Rear Admiral Inoguchi, who remained on her bridge to the end. On hearing the news her divisional commander, Admiral Ugaki, aboard *Yamato* confided to his diary, 'This is like losing part of myself. Nothing I can say will justify this loss.'[1] How then did it come about that one of the biggest battleships ever built, displacing some 72,800 metric tons fully loaded[2], protected by 22,895 tons of armour and 11,661 tons of weaponry[3] could be sunk in exchange for the loss to the Americans of a mere eighteen aircraft? The disparity is such that it is no wonder Ugaki could find no way to explain the day's events.

The first part of the explanation must lie in the decision which allowed *Musashi*, *Yamato* and the remnants of the IJN's surface fleet to be placed in such an invidious position that they were exposed to massive American air attack, with only their own gun defences to rely on at a time when the deadly effect of air bombing and torpedo attacks had long been apparent. Earlier in the war these ships would have been protected by carrier borne fighter aircraft, but the Japanese carrier air arm had been irrevocably broken in the Battle of the Philippine Sea, after which the Americans consolidated their hold on the Marianas islands. The only major units of the IJN Combined Fleet left intact were the battleships, and if the Americans decided to attack the Philippines it was planned that they would be committed to their defence, however slender the chances of success. This was neither desperation nor a desire to gratify the Japanese propensity for self-sacrifice. Admiral Toyoda, the Commander in Chief of the Combined Fleet, explained the reasoning succinctly after the war: 'Should we lose the Philippines operations though the fleet survive, the shipping lanes to the south would be completely cut off, so that the fleet, if it returned to Japanese waters, could not obtain its fuel supply. Whilst if it remained in southern waters, it could not receive supplies of arms and ammunition. Thus there would be no sense in saving the fleet at the expense of the Philippines.'[4]

Defend the Philippines!

The decision to gamble over the Philippines having been taken, the Japanese planners considered several strategems which offered a slim chance of avoiding a one sided contest between carrier aircraft and battleships. One was to rely on surprise and hope the fleet was not sighted as it advanced through the tangle of islands. (In fact the Americans were highly sceptical that the Japanese fleet would be committed to battle, and indeed on the day of *Musashi*'s loss over a quarter of the carrier air power in Task Force 38 was absent refuelling for this very reason). However, since Admiral Halsey's total force mustered nearly 1200 war planes he could still field a considerable strike force. US submarines had also been deployed along all the likely routes the Japanese might take and any hope of surprise was shattered in the early hours of 23rd October when two of these, *Darter* and *Dace*, sank two heavy cruisers[5] and disabled a third[6] while at the same time alerting Halsey to the danger.

The Japanese, however, had not been naive enough to rely solely on surprise and part of their overall plan designated Sho I, Operation Victory, was that the navy's remaining carriers, albeit without aircraft, would try to lure Task Force 38 away from the advancing surface fleet, even at the cost of their own destruction. History shows that this extreme solution bore fruit in the end, but for most of the day of 24 October Halsey was unaware of this force's presence, so that his attention was concentrated on the Japanese surface fleet.

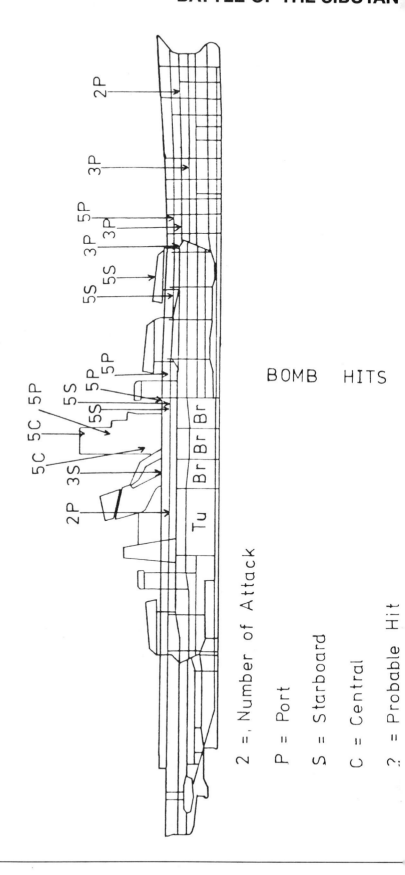

SKETCHES OF DAMAGE TO BAT
BATTLE OF THE SIBUYAN

BOMB HITS

2 = Number of Attack
P = Port
S = Starboard
C = Central
? = Probable Hit

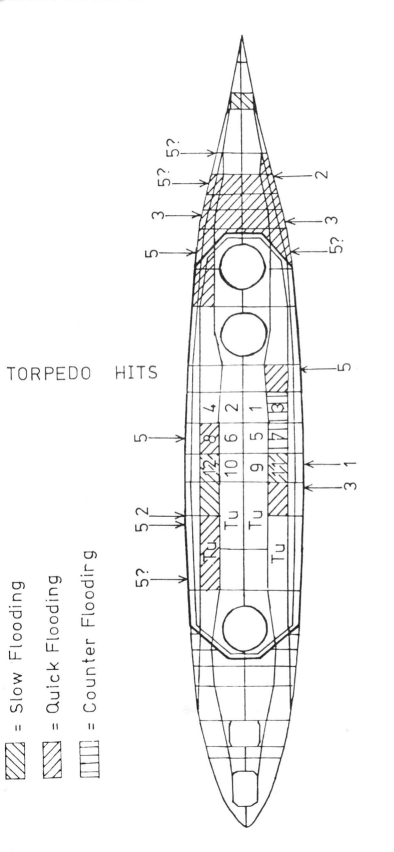

BATTLESHIP MUSASHI DURING
SIBUYAN SEA 24 OCTOBER 1944

TORPEDO HITS

▨ = Slow Flooding
▨ = Quick Flooding
☰ = Counter Flooding

Finally, although the Japanese carrier air arm had been shattered, the Philippines were still controlled by Japan so that in local waters, including the Sibuyan Sea, land based aircraft should have been available to provide the fleet with a combat air patrol. In reality it was left virtually naked[7] as a deliberate decision by local naval air commanders ashore.[8] This was because the standard of Japanese airmanship had so declined that any fighters sent up would provide virtually no deterrent or defence against the veteran American pilots.[9] Instead, the available naval air strength was deployed throughout the morning of 24 October against some of the US carriers where the threat they posed could not be ignored. They sank one carrier,[10] delayed the launching of air strikes from Task Group 38.3 and forced some of these to take off without their torpedoes and armour piercing bombs. Although Morison was scathing about this tactic,[11] given the low quality of the available IJN aircrew it did make an indirect contribution, and the naval air arm could not realistically have done more.

Anti-Aircraft Defences

If the ships could not avoid hostile aircraft, and the weather did not favour them, there was still much they could do to defend themselves. Anti-aircraft gunnery was the principal means of defence left to the fleet and in the event it proved to be inadequate despite considerable efforts to increase their number and effectiveness. As a result of past experience the best formation for air defence was considered to be circular and in the Sibuyan Sea this was duly adopted with the destroyers forming a thin outer shield[12] with the heavy ships in the centre.[13] The theory was that anti-aircraft fire could then be directed to any threatened sector with equal facility. Long range defence relied primarily on the twin 12.7cm (5in) mountings, 24 of which were fitted in *Yamato* and 12 in *Musashi*.[14] Each gun could fire 14 rounds a minute with a fresh gun crew to a ceiling of 31,000 feet, the barrage of each ship bursting in a different colour to assist the director control personnel. Lacking the proximity fuse, fire was generally concentrated as a barrage at the estimated height of the aircraft but this was most effective against high level bombers. Against carrier attack aircraft which were more elusive, short range weapons were handier.

The principal light weapon was the 25mm (1in) cannon based on a Hotchkiss design and as the war progressed the number of these carried multiplied, in part because there was little else available. By this time destroyers carried up to 30 or 40, the heavy cruisers up to 90 and

Yamato 152. *Musashi* had started her career with 24 but by this time she carried 130, 105 in 35 triple mountings and 25 singles. Apart from the problem of ammunition supply for such an enormous number of additional weapons, their main weakness was simply that despite a rate of fire of 220 rounds a minute, and a maximum ceiling of 10,000ft, the individual shells lacked the punch to destroy the sturdily built US carrier aircraft. She was also equipped with four 13mm (½in) machine guns mounted on the forward bridge tower to deter dive-bombers, but despite a rate of fire of 450 rounds a minute this was a puny last line of defence.

Efforts were also made to utilise the battleships' main armament in the anti-aircraft role and both *Yamato* and *Musashi* carried 46cm (18.1in) San Shiki or Beehive shells. A time fuse was set before firing and on detonation each shell released 6000 20mm (0.8in) steel balls in a most spectacular display. In reality it proved impossible to explode these in the midst of attacking formations largely because the latter dispersed before coming into range, and the main armament was incapable of tracking single aircraft. Firing also subjected the exposed anti-aircraft gunners on the single 25mm mounts to severe blast effects. The overpressure from firing a full three gun turret of the main armament was 7kg cm^2, 15m (16.5yds) from the muzzles, and since it was considered that 1.16kg cm^2 could render a man temporarily unconcious, use of the main battery meant the loss of the numerous single 25mm mounts on the flying deck.[15] In the event Inoguchi did not use the main armament until late in the day by which time some 75% of her anti-aircraft mountings had been destroyed. In practice only one full salvo was fired because a shell jammed during loading in one of the guns of the forward turret.[16]

There was also much debate on how the fleet should manoeuvre under air attack. If each ship was allowed freedom of action to avoid bombs and torpedo attacks, apart from the risk of collision, the efficiency of the combined anti-aircraft defence was lost as the gunners became disorientated. Consequently, it was decided that every ship should stay in formation to maximise their firepower, and during the first attack all copied the flagship *Yamato*'s slow turn to port. However, once it became clear that the guns could neither deter nor destroy the hostile aircraft this policy was abandoned. The *Yamato* class was particularly manoeuvrable due to their great beam and comparative lack of length. However, since the US aircraft co-ordinated their assaults, it was to become a case of hit limitation rather than avoidance.

So it was that *Musashi*'s survival ultimately rested on her own robust, passive defences. Before attempting to catalogue the numerous hits she suffered before foundering it is worth remembering that there can be no exact record of the day's events. At the end of the war all official Japanese records relating to her loss were destroyed, and the following is based primarily on the recollections and notes of her Executive Officer[17] Captain Kato, and her Chief Engineer Captain Nakamura[18]. Both should have been in a position[19] to acquire an overall picture of the situation, and yet in some areas where their evidence can be corroborated from other sources both American and Japanese, there are discrepancies even down to the number and timings of attacks. These eye witnesses, particularly after such a protracted and harrowing day, would inevitably have been physically exhausted and disorientated, although by consulting various sources educated guesses can be made to assess the accuracy of their evidence. The exact sequence of events will never be established[20].

Ship versus Aircraft

The first US aircraft were detected at 0810hrs by *Yamato*'s primitive Type 13 air search radar but the first attack by 45 aircraft from Task Group 38.2 did not develop until 1026hrs when the anti-aircraft guns opened fire. Just under half the attackers were fighters and finding no airborne opposition they, like their successors throughout the day, flew in with the bombers to help distract the anti-aircraft gunners and strafe the ships' upperworks. Given their size, *Yamato* and *Musashi* were inevitably singled out for special attention in what was a long drawn out and determined assault. The latter received her first damage when two 454kg (1000lb) bombs exploded close alongside causing splinters which punctured two small peak tanks right forward at frame 20. Since her hull was unarmoured in this region both flooded, though this had only a trivial affect on her trim. Of much greater concern was a torpedo which struck her centrally on the starboard side in the vicinity of frame 130. This carried a 272kg (600lb) torpex warhead which had an explosive force well in excess of that for which her designers had allowed, and her main anti-torpedo defences were breached. The air-filled void outboard of the main armoured bulkhead failed to dissipate the force of the explosion, and the unsatisfactory joint where the main belt, 400mm (15.8 inches) thick, joined the 200mm (7.9in) tapering plate below, sheared pushing both plates inwards and causing small holes in the two residual longitudinal watertight bulkheads.[21] However, slow flooding into No 11 fireroom was contained by the pumps and the resultant 3° list to starboard was reduced to 1° by counterflooding eight to ten anti-torpedo voids on the port side.[22] Overall, apart from some loss of buoyancy, her fighting condition was unimpaired.

The second attack, again from Task Group 38.2, and consisting of 35 aircraft attacked at 1207hrs[23] and lasted no more than four or five minutes. *Musashi*, despite greater freedom of manoeuvre, was not so lucky this time. Whilst splinters damaged the aircraft crane on the stern, two 454kg (1000lb) semi-armour piercing bombs dropped by Helldivers diving vertically from 15,000ft hit her. One failed to explode right forward on the port bow at frame 15 and passed out beneath her bow flare without causing any flooding. The second struck her more squarely six to eight feet to port of the funnel at frame 138. It penetrated at least two decks before detonating; whether it encountered the 200mm (7.9 inch) main armoured deck is unclear. There was certainly damage in the port inner engine-room below because it began to fill with superheated steam and was quickly abandoned, leaving the vessel with only three shafts on which revolutions were increased in an effort to maintain 25 knots. Some sources have suggested that bomb fragments were responsible for severing the steam lines though this is unlikely since the armour was designed to withstand a 1000kg (2205lb) armour piercing bomb falling at terminal velocity. More likely is that the concussion of the explosion caused fractures in the turbine room, which in turn would suggest detonation was very close to, if not on, the main armoured deck. Whatever the truth, fires also broke out at the entrance to boiler room 12 but they were soon extinguished. More irritatingly the ship's steam siren was also damaged and continued to wail intermittently until she sank.

At the same time the ship was also shaken by at least two torpedo explosions.[24] One hit her on the port side near frame 143 close to the bulkhead separating the port outboard engineroom and the port hydraulic machinery space. Once again leaks occurred, principally around the rivets securing the innermost bulkhead, and slow flooding followed in both compartments although it remained manageable for some hours. No port list developed because of the other hit forward outside the armoured citadel at frame 60 on the starboard side. Considerable flooding followed, filling several large storerooms and although the overall effect of both blows was to keep her on an even keel her trim had deteriorated to 2m (6.5ft) below normal.[25]

Despite the loss of buoyancy reserves, *Musashi* was not critically damaged although on three shafts she could not make more than 22 knots, three less than the rest of the formation. In consequence she started to drop behind but once the attack was over, speed was reduced to 20 knots so the damaged battleship soon regained her station of comparative safety. A prudent commander might have ordered *Musashi* to withdraw though given the desperate nature of the whole operation this was considered pointless. From a more cynical point of view, and one of which Ugaki appeared unaware, the retention of the battleship had merit in that there were two big targets which would undoubtedly continue to attract US aircraft to the benefit of their smaller less resilient companions.

The third assault of 35 aircraft from Task Group 38.3 came within range at 1331hrs. They took their time and seeing that *Musashi* was clearly damaged she suffered heavy and concerted attacks. Four bombs hit her, three quite close together forward of No 1 main turret at frames 45, 65 and 70. Close to the turret the flying deck was lightly armoured between 55mm and 35mm (2-1.5in) to counter the effects of blast but the bombs were too heavy and smashed through it and the deck below before exploding in unoccupied crew accommodation spaces. No fires occurred which might seem surprising given the notoriously combustible nature of Japanese warships, but this can best be explained by flooding resulting from adjacent torpedo hits. The last bomb fell to starboard outboard from the funnel at frame 134 and exploded on impact. Either it was a 227kg (500lb) general purpose bomb designed to detonate in this way, or equally plausibly it could have landed on a two-ton triple 25mm mounting which stopped further penetration. Whatever the cause, it shattered the massed anti-aircraft batteries in its vicinity and caused heavy casualties.

The Americans also managed to put at least three torpedoes into the embattled vessel. Two went into her bow, one on each side at frame 70. With no torpedo protection, flooding was massive and spread right across her hull on the middle deck[26] from the forward edge of the citadel to frame 54, the water finding few barriers due to bomb damage. To onlookers the damage was obvious; coupled with the earlier blow to her starboard bow the plating had been forced outward causing a huge column of water to be thrown up as she continued to drive forward. Another torpedo hit her to starboard at frame 138 and since her main defences had already been ruptured at frame 130, flooding into the starboard hydraulic machinery space was immediate and it had to be sealed off and abandoned. A fourth torpedo hit near frame 110 to starboard has been reported although once again the lack of flooding reports and an overall list of only 2° to starboard suggests that no hit occurred.

The bombs had caused no serious damage and the list was reduced by counter-flooding the remaining outboard port voids, but her general condition was now serious due to her trim. From a freeboard of 10m (32.8ft) right forward it was now down to 6m (19.6ft) and she was drawing over 15m (50ft). Her best speed was down to 16 knots and even after the main formation had reduced speed she continued to fall steadily behind. Fears were voiced that even at her reduced speed she might suddenly plunge by the bow, helped by her own colossal momentum, and Inoguchi cautiously reduced speed further to just 12 knots.[27] Her lack of speed proved beneficial temporarily because the fourth air strike of the day at around 1430hrs ignored her since she seemed to be out of the battle.

Musashi *Retires to the West*

This proved to be the case because she was ordered to make for the west with two escorting destroyers. This was a sensible decision for she lacked the speed and manoeuvrability to be anything more than a liability. However, despite six bomb and six torpedo hits she was not in any real danger of sinking and there was no progressive flooding. The American team investigating her loss believed that in the two hours of peace that elapsed before the final fatal attack, more should have been done to regain buoyancy by pumping out the voids on both sides of the ship used earlier for counterflooding. Each was equipped with a steam ejector capable of lifting 200 tons per hour, but although this would have helped her trim, her final demise would only have been delayed not avoided.

The Final Phase

The fifth and fatal attack came from Task Group 38.4 and consisted of 65 aircraft. It might be thought that *Musashi* would again have been left alone but by a quirk of fate the main Japanese formation had temporarily retreated to the west, and when the attackers closed shortly after 1500hrs over half ignored the organised main body and concentrated on *Musashi*. She was certainly a most attractive target; her anti-aircraft defences largely silenced, steaming slowly northeast at 12 knots and unable to manoeuvre rapidly. The importance of these elements to a ship's survival were about to be graphically demonstrated.

Within five minutes she received ten 454kg (1000lb) semi-armour piercing bombs, dropped by Helldivers coming in over her bow and concentrating on her towering main superstructure. Working aft one fell in the tangle of wreckage in front of No 1 main turret, to port at frame 62. The next hit the roof of the turret at frame 75 and exploded on impact, leaving a soup bowl sized score about one millimetre deep in the 270mm (10.6in) thick toughened steel plate.[28] Another at frame 79 fell to starboard of the turret and penetrated two decks before exploding against the 230mm (9in) sloping protective bulkhead of the main citadel. There was no damage beneath, although the wardroom above was devastated. The fourth and fifth fell close together and detonated on the flying deck at frame 115 slightly to starboard in the gap between the main superstructure and the forward triple 15.5cm (6.1in) secondary gun turret. Two more fell to port, one at frame 115 again, the other slightly forward at frame 108. They penetrated two decks exploding on the 200mm (7.9in) armoured deck, the damage again being confined to the spaces above.

The eighth bomb also fell to port at frame 120 on the side of the superstructure, exploding on impact, wrecking anti-aircraft guns and damaging the air intakes to the boiler rooms below. The penultimate bomb could be said to have hit the bullseye as it landed on top of the main tower at frame 120, destroying the main gun director and rangefinder and damaging the bridge and operations room. Seventy-eight were killed and wounded, the Captain being amongst the latter with a shrapnel wound in the right shoulder. The last bomb fell on the centreline at frame 127 exploding on contact at 02 deck level[29] inside the ring of anti-aircraft batteries causing only minor damage. In total, although dramatic in effect this damage remained relatively superficial since neither power nor buoyancy, the vital ingredients to her survival, were directly affected.

However, it must be remembered that as the ship reeled beneath these blows her flanks were being torn open by at least four torpedoes, and she may have been hit by as many as ten. Not surprisingly, information on this attack is hard to come by, due to its ferocity and the fact that centralised control of the damage repair teams started to break down. One confirmed hit was on the port side at frame 75 in the way of No 1 turret magazine. The 340mm (13.4in) thick angled bulkhead enclosing the turret was not immediately adjacent to the ship's side, and given the vital importance of space to dissipate the force of an explosion, this should have kept the citadel intact. However, flooding was soon reported in the two lowest levels of the magazine, and the most likely explanation is that the earlier hit at frame 70 together with other bomb damage had already seriously weakened the strength and watertight integrity of the citadel. Another hit, also to port was at frame 125 and caused rapid flooding and the abandonment of No 8 fireroom,

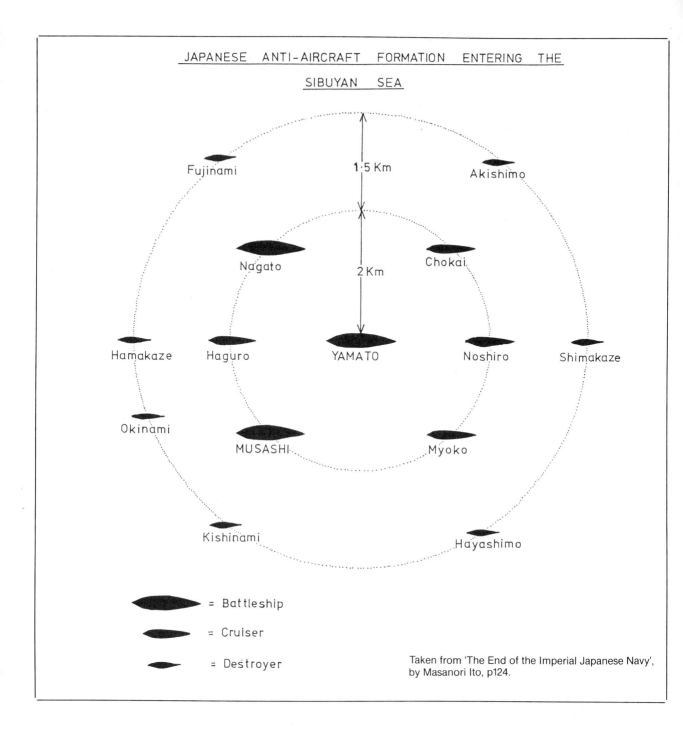

Taken from 'The End of the Imperial Japanese Navy', by Masanori Ito, p124.

whilst that into No 12 fireroom remained controllable. Such radical failure was surely the result of the accumulated blows to which she had been subjected.

The third hit to port was at frame 145, just two frames aft of an earlier one, and there was immediate flooding in the port outboard engine room although, mercifully, most of the personnel managed to escape. These three blows left the ship with only her two starboard shafts and a severe list to port. This list was countered by a hit to starboard at frame 105 which breached the main defences and allowed water to leak into the anti-aircraft magazines on two levels forward of No 3 fireroom.

The end approaches

Of the remaining six possible hits, two were thought to have been to port at frame 140 and were said by eye witnesses not to have exploded. Captain Kato also noted two to port at frames 40 and 60 and one to starboard at frame 80. Since there was already severe flooding in these areas the lack of flooding reports is not surprising and it is possible that they did occur. A last hit was noted to port at frame 165 in the way of the main after magazine but he had no reports of flooding, though other sources do suggest this did happen, and the report is supported by the severity of her list to port.

At the conclusion of the attack this had climbed to 10° - 12° and her trim forward was lower by another two metres. In this condition and on two shafts she could only make six knots and this was not enough to even give her steerage way. On hearing this news Ugaki ordered her to try to beach herself on nearby Sibuyan Island to the south but this was beyond her.

The end of her struggle approached inexorably; her crew were highly trained even for the Imperial Navy and they had spent much time in damage control exercises, but they were unable to correct either her trim or list. Efforts made to flood storerooms on her starboard quarter failed since they were not fitted with seacocks and when hoses were laboriously run from the firemain there was insufficient water pressure left in the system. Captain Nakanmura, the Chief Engineer, on his own authority then ordered the deliberate flooding of the starboard outboard firerooms, Nos 3, 7 and 11. This drastic step held the list at 12° although it did nothing to remove it and on her remaining screw she was barely able to move.

This wisdom of this decision scarcely mattered because the flooding could not be contained and no watertight boundary was ever re-established. By 1630hrs an hour after the last attack she was visibly lower in the water and one by one her remaining boiler rooms had to be abandoned, the last being evacuated by 1800hrs. By 1900hrs the water was lapping round the portside of No 1 turret and the list had started to increase again, reaching 15°. Only then was the decision taken to abandon ship and by 1920hrs the list had doubled to 30°.[31] Within half an hour she had capsized and disappeared.

For all the faults in her design her defences had performed certainly to the limits her designers would have expected, and it proved if nothing else that greater size permitted stronger defences without penalising speed or fire power. She had been struck by at least 16 bombs, the great majority designed for attacking ships, and yet apart from having to abandon one engine room due to an escape of superheated steam, none had penetrated her central armoured citadel or threatened her survival. Unless bombs of great size were used the battleship had little to fear from that quarter; it was torpedoes which really sank *Musashi*, or to be more precise, some 14 of them. There could never be much margin for safety given the severe constraints under which designers laboured and with the unforseen arrival of torpex, which doubled the power of any torpedo detonation. Flooding inside the main armoured box was certain, unsatisfactory joint in the main bulkhead or not. With multiple hits sometimes close together any battleship would have foundered.

There has been criticism that because only 53.5% of her length was protected this left too great an area exposed. In *Musashi*'s case this led directly to her trim problems. However, if the armour had been spread wider it would have had to be thinner and so less capable of resisting damage, which might in turn have led to more wide-spread flooding much earlier.[32] Compromises of necessity have their drawbacks and that chosen was as good as any other.

Such was the strength of the vessel that after the battle the American analysts were dismayed at the resilience of the *Yamato* class, in part because they were quite unaware of their size.[33] However, they did observe that capsizing would be more likely, and would come that much sooner, if torpedo attacks were concentrated on one side. Surprisingly *Yamato* survived Operation Victory and in the spring of 1945 this lesson was to be put to good practice.

Footnotes

1. Battle for Leyte Gulf, by E P Hoyt, p 156.
2. A metric ton equates to 2,204lbs and an English ton 2,240lbs. Thus in English tons she displaced 71,650 tons. In this article all tons are metric.
3. Calculations and assumptions for these figures can be found in Battleship Design and Development, by N Freidman, pp 170-171.
4. Battle for Leyte Gulf, by Adrian Stewart, pp 15-16.
5. *Atago* and *Maya*.
6. *Takao* (by *Darter*).
7. There were according to the Americans never more than four Japanese fighters over the fleet and these were easily disposed of. History of USN Naval Ops in WW II, Vol XII Morison, p 184.
8. Even at this stage of the war there was minimal co-operation between the Japanese Army and Navy airforces, and the former took no part in these operations.
9. Their lot would have been made even more perilous because the Japanese gunners not unnaturally took all aircraft as American and shot at any that came close.
10. CVL 23 Princeton.
11. op cit. Morison, p 187.
12. By comparison with Allied navies this fleet was very short of escorts, largely because they had been diverted to trade protection during which they suffered heavy losses.
13. Some sources state that the arrangement in the diagram was only adopted after the first attack, rather than from the outset.
14. The discrepancy between the two sister ships was due to the fact that the original beam mounted secondary low angle armament had been removed to save weight and space for additional anti-aircraft guns but there were none available for *Musashi* at the time and they were never fitted subsequently.
15. The flying deck was the term used by the IJN for what would be the upper deck in the RN and the main deck in the USN.
16. The blast problem did have a positive side; all the ship's boats were stored aft under cover and that left the firing arcs of the anti-aircraft guns remarkably uncluttered.
17. The Executive Officer is the Second in Command.
18. This information comes from US Technical Mission to Japan, No 2-06-2 Reports of Damage to IJN Warships, pp 17-21.
19. Captain Kato spent the day in the main superstructure and he swam clear as the vessel turned turtle. Captain Nakamura was stationed in the inner enginerooms and when she rolled over he ran up the side and over the bilge keel before being thrown off.
20. In the most recent definitive account of this event found in Axis and Neutral Battleships in World War II by Dulin and Garzke, they use the times quoted in S-06-2. These cannot be supported from any other source, including *Yamato*'s Action Summary for that day found in Report AR-209-78.
21. For a fuller account of this system see the article in *Warship*, No 41 by the author.
22. Given their great beam of 38.9mt (127.5ft) at its maximum, flooding of outboard compartments was bound to cause list problems. Thus there was great stress on effective counterflooding equipment and procedures, during damage control training.
23. Nakamura and Kato believed there was another attack at 1140hrs but there is no evidence to support this. The damage they ascribe to it is believed to have occurred in the 1210hrs attack.
24. According to both officers she suffered two more hits to port near frames 80 and 110 outside the central citadel, but the absence of any flooding reports, or list to port, casts doubt on the accuracy of these claims.
25. At the start of this operation her draft had been 10.9m (36ft). Before the battle it had reduced to 10.5m (34.5ft) since she had burned 1000 tons of fuel oil and transferred 800 tons to the escorting destroyers.
26. The Middle deck was two decks below the flying or upper deck.
27. It had been calculated by her designers that she could operate with her trim reduced to 4.5m (14.8ft) at the bow. Thus Inoguchi may have erred on the side of caution. USNIP, October 1952, p 1110.
28. Garske and Dulin state No 2 turret was hit in this raid and that the hit on No 1 occurred during the first attack. No clear answer would appear to be possible.
29. Above the flying deck or upper deck each level was generally listed in ascending order starting at 01 upwards. Thus 02 was two decks above the upper deck.
30. op cit, Garzke and Dulin, p 72.
31. Stability was believed to become questionable at 20° of list.
32. There were also criticisms that the compartments in the bow area lacked sufficient subdivision to limit flooding. Whilst true, proposals to improve it were rejected because of the weight penalty. In *Musashi*'s case bomb damage would have negated most of the benefit.
33. Secrecy was considered an essential part of their construction and the Japanese went to quite extraordinary lengths to ensure it. They were completely successful and the truth only emerged after Japan's surrender in August 1945.

LANDING CRAFT THROUGH THE AGES

PART I
Brian Friend

Photograph of a model in the National Maritime Museum of a flat bottomed landing boat with crew and soldiers embarked.

A reconstruction of a Roman vessel of the second century. The flat bottom suitable for grounding is very evident from this drawing.

During the six years between September 1939 and September 1945 over 74,000 landing ships, craft and amphibious vehicles were built or converted by the combatant powers. They came to be known collectively under the general name of Amphibious Warfare Vessels (American Forces) or Combined Operations Types (British Forces). The Axis Powers assigned similar designations to their vessels, although in many cases their units were built, manned and operated by the Army.

These vessels ranged in size from the 35,739 gross tonnes of the mercantile manned Landing Ship Infantry (Large), LSI (L) *Mauretania*, the pre-war ocean passenger liner, to the diminutive Landing Craft Rubber (Small), LCR (S), a tiny seven-man inflatable rubber boat, propelled by the paddles of its occupants, and designed for covert raiding operations.

Their official designations ranged through the alphabet from the AGC (Amphibious Force Flagship) of The United States Navy, to the 'Z' lighters, built by the British in India, which were capable of being shipped in sections overseas, and then assembled by local unskilled labour. So numerous were these craft that most were given serial numbers, but not named. Also, many merchant ships that carried landing craft, and were used in combined operations, never received an official designation.

The majority of this enormous armada were conceived, designed and built in an incredibly short time. Of the major combatants, only Japan had a large amphibious landing capability prior to World War II. The Royal Navy possessed a few Motor Landing Craft, (MLC), built under the exigencies of The Landing Craft Committee in 1926 and 1929. (This committee had been formed in the early 1920s on the recommendation of the first of the inter-service Staff College meetings). These craft were capable of carrying 100 troops and one or two vehicles, but were very primitive and had only been used for trials and training, although some of the later versions were used, and lost, in the Norwegian campaign of 1940. There were also a small number of prototype craft of new designs in service, with some 18 LCA, 12 LCM and 2 LCS (M) of the new designs on order. Plans had been prepared to convert certain selected ships into assault transports but the drawback was that the Admiralty would have to requisition them and then be persuaded to release them for conversion. The United States Navy had a converted destroyer, the USS *Manley* APD 17, the world's first high speed raiding transport, carrying four Higgins 'R' boats and a company of Marines. They also had two commissioned transports, the *Henderson* and *Chaumont*, but both were employed on routine trooping duties. A powered tank landing lighter had been developed during the 1930s, and through a series of fleet landing exercises starting in 1922, but held annually from 1935 to 1941, the doctrine of The Fleet Marine Force as the amphibious assault arm of the US Navy was established. In 1933, the formation of

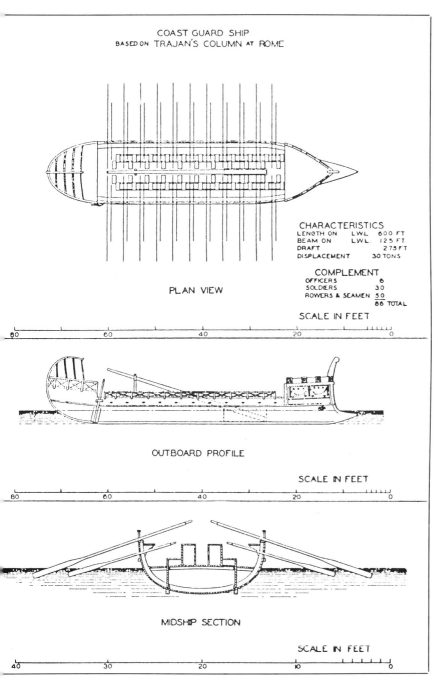

delivering the landing force to the correct beach.

It is a widely held view that *Shinshu Maru* was the world's first specially designed and built landing ship, and that the techniques and specialised equipment needed for amphibious warfare were only developed during the second World War. This is not the case, for specialised landing craft have been built or improvised for thousands of years, and have been in use since the beginning of recorded warfare.

The Early Mediterranean Experience

The need to transport troops across the seas and land them in full battle-order upon a hostile shore is as old as history itself. The transport systems of the early civilisations in the Mediterranean basin were invariably, with exceptions, waterborne. Their ships were flat bottomed and shallow draft types, intended for landing both men and goods on suitable beaches, either for the night or other business.

At the Siege of Tyre in 332–331 BC, Alexander used his forces in a series of combined operations over a period of seven months to breach the walls and take the city. His assault from land (a mole was built from the shoreline to the city) was coordinated with a seaborne attack by his fleet. Engines of war were set up on the mole and mounted in ships, lashed together in pairs for stability. These were brought up close to the eastern wall, south of the mole, and used to make a breach. The remainder of the fleet was employed to protect the assault craft from counter-attack, clear obstructions, and harass the defenders with various types of missile. Two ships had been modified to carry the assault force, led by Alexander himself: 'Finally the breach was ready and the transports charged in and threw out their gangways to the wall. Alexander was the first to cross and enter the breach and lead his storming party to the palace, but the city was entered from elsewhere, also.'

Other squadrons had forced the defensive 'chains' and overwhelmed the Tyrian Fleet. The victorious crews had then landed and added their weight to the assault on the city. The siege of Tyre is regarded as one of Alexander's greatest exploits, and was possible only because of his command of the sea and application of amphibious warfare techniques. 'It was largely owing to the small draught of the ships that this could be done. The men on board could be brought so close to the shore that they could step overboard in two or three feet of water, while their supporters, remaining on board, had the

The Marine Corps Equipment Board had established a body of professionals whose terms of reference instructed them to 'devote their entire time to the study and development of material suitable for the use of troops in amphibious warfare.' According to this author's research, the Italians and Germans had no specialised landing vessels whatsoever.

The Japanese, with their experience of landing operations in China, had the largest number of specialised landing vessels in service. These were the *Dai Hatsu* type of small landing craft. Several hundred of them were observed in The Cut-Off at Shanghai in 1937 by Rear Admiral L H Mund (then a Captain) whilst salvaging a fellow officer's houseboat during Japanese operations in the city. They also had the *Shinshu Maru*, a specially built landing ship of 8100 tonnes, designed to launch the *Dai Hatsu* pattern boat, in pairs over a ramp at the stern, with loading ports amidships for passing stores and men into the craft. These vessels, unlike their allied counterparts, were built and manned by the Japanese Army. In Japanese service doctrine, the responsibility for the actual landing rested with the Army, the Navy being charged with the task of transporting the expedition over the oceans and

35

The USS *Manley* seen here as AG 28, after her first limited conversion to experimental destroyer transport. Note the Whaleboats, the four sets of gravity davits, and the small luffing davits aft. Her dimensions and performance were the same as the famous Four Stackers of World War I, of which she was one of the first built.

advantage of a higher position to shoot down on the enemy on the beach.'

The Siege of Tyre was probably not the first time that ships had been modified for an amphibious assault, and it was certainly not the last. Thirty years later, Demetrius laid siege to the city of Rhodes. Ships were once again adapted for use: 'He built tortoises to give shelter against the enemy missiles and mounted each on two ships fastened together. He also mounted two battering towers, four stories high, in the same way, and then built a floating boom of heavy timbers to protect the ships carrying the engines from rams. He also assembled numbers of smaller craft and decked them over in pairs, so that they might carry catapults and Cretan archers.'

All this, however, was to no avail. The city held out, and after a year of siege, peace was signed in 304BC and Demetrius sailed away.

The Romans and After

The Romans both built and improvised specialised landing ships and craft. They used fleets and armies acting together to expand and then keep their empire. Julius Caesar invaded Britain in 55BC by crossing the Channel. This same operation was repeated, with occupation in mind, in later years by his successors, and over 800 years after Alexander's siege of Tyre: 'Belisarius attacked the Gothic garrison of Palermo from the sea in AD535. He was bothered by the defenders in the ramparts of the harbour, who overlooked the ships. He put his archers into boats, hoisted the boats to the mastheads, caused the archers to fire down on to the defenders, sailed with impunity into the inner harbour and captured the city.'

A case of improvised firepower support.

The Vikings turned raiding into an art form, using their magnificant long ships, developed over the years, with devastating effect.

William the Conqueror built a fleet with which to invade England. We do not know the exact composition of his army but it is generally accepted that it must have totalled about 8000 men, including 2000 mounted knights. Norman boats of that period had a capacity of about 20 armed men, and applying the general

The USS *Manley* now designated APD 1 after her full conversion to destroyer transport. Two stacks and boilers have been removed; additional accommodation and fuel stowage added. Four Higgins *Eureka*s at her davits, along with a company of marines, was now her main armament. The photograph shows her just after her conversion in the Cape Cod canal in 1940. Thirty-five other flush deckers, and nearly 100 DEs were converted to destroyer transports (APD), by the USN during World War II.

rule that one horse equals five men: 'about 1000 boats had to be constructed to carry the army and necessary non-combatants, 2000 or so horses and a certain quantity of provisions and war materials.'

It is inconceivable that if boats were being built to carry horses, and 250 would have been needed to ship William's estimated 2000, they would not have been provided with boards for the horses to stand upon during the crossing, and with ramps or tackle with which to disembark the animals when the invasion fleet beached at Pevensey.

The Crusades

The Crusades underlined the need for horse transports and special vessels were built at the ports of embarkation. Each war horse was suspended in a special sling whilst at sea. On arrival at the landing place, usually a gently shelving beach, the transports were beached stern first and ramps lowered over the stern, down which the war horses were led. In the fourth crusade in 1204, during the second assault on Constantinople, the attack was made directly from ships against the walls that bordered on the Golden Horn. After their first attacks were repulsed, the Venetians and French spent two days making repairs and preparing for a further assault. '40 ships (uscicri) were joined in pairs, each with flying bridges so that each pair of ships could overpower the men in the tower opposed to them.' Siege engines and catapults had been mounted in the bows of warships: 'The main line of ships had wickerwork on their sides to break the force of stone missiles. There were hogsheads of vinegar to extinguish Greek fire, and there was further protection to the ships from skins, iron plates and baled wool.'

The flying or boarding bridges 'were built upon the antennae (or yards) with rope side guards whose sides were protected by leather.'

The attack finally succeeded when two ships lashed together, the *Pilgrim* and the *Paradise*, assaulted a tower, landed a bridge on its top: 'and a bold Venetian and a Knight of France entered the tower.'

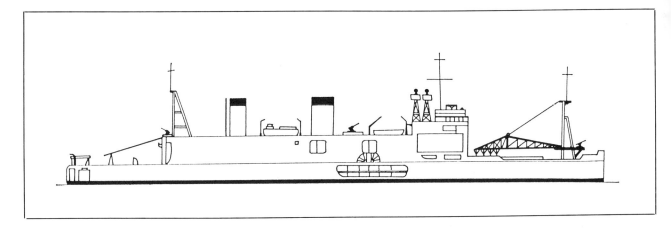

The *Shinshu Maru* as she appeared in 1937 before major alterations Displacement; 12,000 tons. Dimensions: 492ft × 72ft, 2in × 26ft, 9in. Machinery; 2 shaft geared turbines 800HP = 19knots: Capacity; 20 Dai Hatsu landing craft and troops.

Amphibious warfare was conducted by the Turks against the Knights of St John on the islands of Rhodes and Malta, by Drake at Cadiz, and by the Armada against England.

The Duke of Alba, who had been named as the first commander of the Armada (he was to die in 1582 before it sailed), listed amongst his requirements a need for 200 flat 'barcas', 'or barges of a type that had been found useful in the Spanish conquest of the Portuguese Azores'.

In 1594, Sir James Lancaster outfitted an expedition to Brazil. It was well organised. 'The well governed and prosperous voyage of Mr James Lancaster, began with three ships and a galley-frigate from London in October, 1594, and intended for Fernambuck in Brazil'.

Carried onboard the ship as part of the military stores were the frame and timbers for a galley, 'of purpose to land men in the country of Brazil'.

Operations in the Baltic and the French Wars

During the seventeenth century Gustavus Adolphus II of Sweden made great use of combined operations to expand and consolidate his kingdom around the Baltic at the expense of Russia and Poland. By the middle of the eighteenth century the Royal Navy was building specialised landing craft and adapting ships for amphibious warfare. A tactical doctrine and signal organisational had been worked out and was used during the Seven Years War against France (1756-63), in the American War of Independence (1775-82) and the Napoleonic wars. Flat bottomed boats were used at the taking of Louisburg and Quebec and as always they were in short supply. The study of contemporary sources reveals that the degree of professionalism achieved by the British during these operations was very high. The organisation, equipment, signal procedures, fire support ships and above all the use of flat bottomed boats were firmly established. These craft were built after the failure of the Rochefort expedition of 1757, where the need for a specialised form of landing boat was one of the many lessons learnt. 'Two different sizes of Flat-bottomed boats were developed. The larger type was 36ft long, 10ft, 2in breadth, 2ft, 11in in depth amidships between keel and gunwhale, and was equipped with tholepins and thwarts for 20 oarsman. The smaller type of boat was 30ft long, 9ft, 9in wide, 2ft, 11in deep amidships, and was fitted for 16 oarsmen.'

Both types were of similar construction. 'They were flat-bottomed, clinker-built, with bluff bows, and steered by a detachable rudder and tiller. Two ring bolts were mounted on the keel inside each craft in order to hoist the boat in and out of the water.'

A list of equipment to be carried in the boats can be seen in the Public Record office, and individual officers added to the list according to their own ideas and experience. Other types of craft were built and a number of ships modified for amphibious work. 'In 1776 at Staten Island the Army built a number of craft capable of carrying 200 soldiers and having ramps mounted on their bows for unloading cannon.'

Ships were fitted with heavy gear to hoist the flat bottomed boats in and out. A merchant ship was specially converted during the American War of Independence to mount sixteen 24pdrs for the sole purpose of shore bombardment. The doctrine of how these forces were used was firmly established, and preceded the large amphibious operations of World War II by nearly 200 years. Standing orders laid down that the craft were to be guided to the beach by larger warships, recognised the need for a proper reconnaissance of the landing area, and called for adequate fire support before and during the assault. The orders also laid down that all ranks taking part were to be thoroughly briefed and exercised beforehand. As we can see, the problems had been studied in depth and solutions provided in the mid eighteenth century.

Napoleon and the Channel

Napoleon, preparing for his assault on England from his camp at Boulogne, built hundreds of 'barges': 'prams (sailing ships) of about 400 tons load displacement (that) could each take 80-120 soldiers with 50 horses and 38 sailors and 10 days rations and forage for all, besides some military supplies and guns, amounting to perhaps 60 tons.'

He also built gunboats of 120 tons carrying 130 men, and several types of smaller craft, all expressly designed for the channel crossing and a subsequent amphibious landing in England.

Throughout the nineteenth century, during that period which saw the major expansion of the British Empire, amphibious operations continued, though specialised craft were seldom built for the task. The American expedition against Mexico in 1847, however, saw the construction of specially designed 'Surf' (landing) boats. 'These were designed by a naval officer and came in three sizes so that they could be stacked for transportation. Each carried approximately forty men. The crew consisted of eight men: a commander, a coxswain, and six oarsmen. Each boat was equipped with a kedge anchor and line for re-entry into the surf and to prevent broaching.'

The Crimean War commenced in earnest when British and French forces made an unopposed amphibious landing on the beaches of Calamita Bay, about 30 miles to the north of Sevastopol and close to the small town of Eupatoria. Elsewhere throughout the globe, soldiers, sailors and marines were landing to rescue missionaries, put down revolts, protect national interests or acquire overseas possessions; the era of colonialism was approaching its high point.

The Imperial Japanese army ship '*Shinshu Maru*' engaged in the landing operations off Woosung, China, in November 1937. 'Dai Hatsu' types of landing boats can just be made out alongside, at the stern and amidships.

FIGHTING MOTOR BOATS OF THE RUSSIAN ARMY
Eastern Europe 1915–20

Dipl-Ing René Greger

Whereas prior to World War I the Russian Government maintained a large force of armed launches on the River Amur in the Far East, no such force existed in Eastern Europe as the frontier between Russia and her potential enemies, Austro-Hungary and Germany did not follow the line of a great river system.

The situation changed dramatically in the summer of 1915 when the retreating Russian armies had to evacuate central Poland, and the German offensive against Riga ground to a halt on the banks of the western Dvina. The front line now ran through the extensive water system formed by the Pinsk and Pripet Marshes. The Czarist generals soon recognised that a fleet of armed launches would be of tremendous value in future operations. Unfortunately, neither the Army nor the Navy possessed suitable craft in any numbers, but undeterred the Army quickly drew up plans for three mobile formations, each to consist of three motor gunboats, six armoured motorboats, six small patrol motorboats, six high speed launches and four small minesweeping cutters.

The next problem was one of procurement. Russia's low productivity and the overstretched boat-building resources of her European allies meant that the smaller craft had to be ordered further abroad, in this case in the USA where contracts were signed for a total of 63 marine engines, 18 armoured patrol boats and 30 small infantry assault motorboats. There remained the question of the motor gunboats and the orders for these were placed with small private yards, the necessary drawings being produced by the Corps of Army Engineers. The *Kopye* Class in service on the Amur were taken as a model for the Retchnaya Kanoners-

RKL Type motor gunboat in service with the White Army on Lake Onega in 1919.

kaya Lodka (*RKL*) as they were known. Offensive operations were planned to commence in 1917, but due to delays in the supply of materials and the shortage of skilled labour (at the front), the target date was not met. The nine motor gunboats were constructed in Finland, at Odessa and at the new yard of Bekker & Co at Reval (Tallin) in Estonia. All were delivered late in 1917, but none saw service on the German front, unlike the small armoured patrol boats built in the USA and delivered via Vladivostock some months earlier but which were not fully commissioned at the time of the October Revolution. These craft, built by Mullins & Co, were known as Bronekater (*BK*).

The *RKL* were taken over by the Army just prior to the Revolution but they had to be manned by personnel from the Baltic Fleet and were concentrated at Petrograd for use in the coastal waters of the Gulf of Finland. After the Revolution they were allocated, in May 1918, to the Red Flotillas on Lakes Ladoga and Onega. Four of them participated in the successful operations against the Finns on Lake Ladoga in June 1919. However, one of the photographs accompanying this article clearly shows an *RKL* flying the St Andrew's Cross used by the White Russians; presumably on Lake Onega sometime in 1919.

According to Soviet sources they were widely employed during the Civil War and saw service on the Volga and in the Caspian Sea. Ultimately, they were returned to the Baltic and served with the Frontier Guard Flotilla. Seven were on strength until 1940 and one saw service in World War II as a sea-going patrol boat. It is believed that the 'K' Type motor gunboats mentioned in the wartime Soviet Naval Annual 'Spravotchnik 1944' were *RKL*s under another name.

Eighteen armoured motorboats were built, four by Bjorneborgs at Borgo in Finland and 14 by Revenski & Gonzevich in their private yard at Odessa. The four Finnish craft were completed in late 1917 and accompanied the Red forces when they left Finland in the spring of 1918. Nothing is known of their history after they arrived in Petrograd until June 1926 when they were all listed on the strength of the Dnieper Flotilla (Soviet-Polish Frontier Watch). It appears that at least some of the 18 were still in service up to

Standard motorboat of the BK Type (BK13) as used by the Soviet Amur Flotilla.

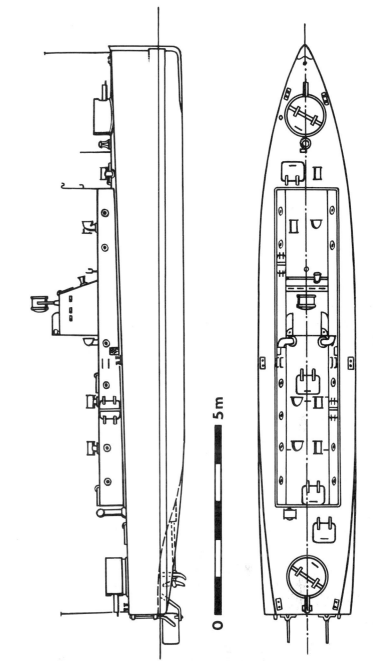

Armoured motorboat as built in Odessa and elsewhere, intended for reconnaissance duties.

	Displacement	Length	Breadth	Draught	Speed	Protection
RKL	24 tons metric	20.4m	3.2m	0.6m	23 km/hr	5-6mm
BK	7 tons metric	9.3m	2.4m	0.6m	20 km/hr	5mm (cabin)
Armoured motorboats	15 tons metric	16.0m	3.1m	0.6m	25 km/hr	6mm

POWERPLANT
RKL Two 75 HP 'Buffalo' petrol engines, one 7 HP auxiliary petrol engine.
BK (USA built) One 'Stirling' petrol engine of 80 or 97 HP (Sources vary).
Armoured motorboats. Two 50 HP 'Stirling' petrol engines.

ARMAMENT
RKL Two 76mm mountain guns, two mg. *BK* One mg originally. *Armoured motorboats* two mg.

Notes: petrol = gasoline.
Observant readers will notice that the data quoted above differs slightly from that given for the armoured motorboats in the author's *Austro-Hungarian Warships of World War I*. Ian Allan, 1976. They were acquired for use as minesweepers.

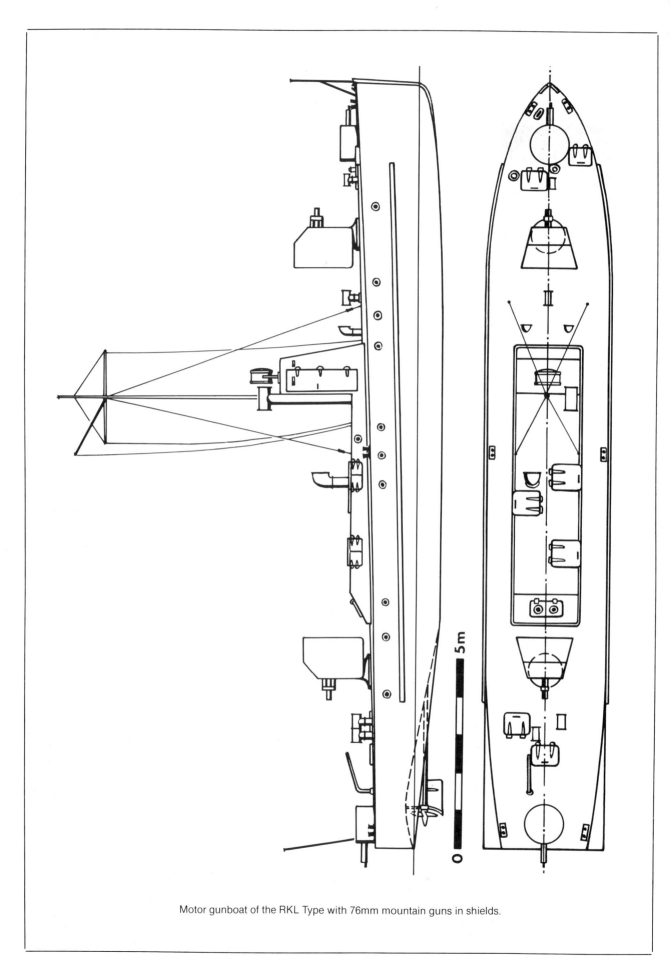

Motor gunboat of the RKL Type with 76mm mountain guns in shields.

Photograph taken at Novisad in 1920. The craft on the right is one of the armoured motorboats built at Odessa and acquired by the Austro-Hungarians.

the outbreak of the war. 'Spravotchnik 1944', although unreliable as a source for technical data, refers to them as the 'N' Class. Those built at Odessa had a rather more exciting history – only two were complete when the Treaty of Brest-Litovsk was signed on 3 March 1918 and the Austro-Germany forces moved eastwards to occupy the Ukraine. Odessa was taken on 13 March by the Central Powers. The two complete boats were evacuated to Cherson, where the Austro-Hungarian monitor *Szamos* seized them some weeks later. As Odessa was occupied by the forces of the Dual Monarchy, all captured enemy property belonged to the state. Being first on the scene the army laid claim to the craft under construction in the yard but were unsuccessful as the status of the Ukraine changed when a puppet government was set up under the auspices of the Central Powers and the Ukraine became an independent state. No longer enemy property, the armoured motorboats had to be purchased, and the Donauflottille of the 'K und K' put in a bid. The asking price was very low and about half of them were close to completion, however there was a snag – a grave shortage of the necessary skills. In fact, the Austro-Hungarians were only able to take possession of four craft; the other eight were still incomplete when the Red Army marched into Odessa in April 1919. Three of the type were at Budapest in November 1918 and subsequently served with the Hungarian River Police, minus their machine-gun turrets. Listed as patrol craft under the names *Honvéd*, *Huszàr* and *Tüzer* they appeared in the naval annuals until the conclusion of World War II although only the last named was on strength in April 1941 when she participated in the short campaign against Yugoslavia. It is open to doubt whether a fourth motorboat named *Hedervár* was of the same type; she served as an auxiliary in the Hungarian service and may have been one of the other types ordered by the Czarist Government in 1916.

Another ex-Czarist armoured motorboat served under the Yugoslavian flat on the Danube until the late 1920s. In the absence of any positive information, it is assumed that she was one of the craft abandoned by the retreating Austro-Hungarians at Novisad, formerly Ujvidek, in November 1918 and subsequently commissioned by the Yugoslavs. According to a Soviet source, the Germans captured several motorboats in Sevastopol in May 1918 which they handed over to the Turks. The prizes should have included two armoured motorboats but Turkish sources do not confirm this.

Many photographs of the type *BK* exist, some of them are believed to have seen service in the northern sector, for at least one fell into German hands on Lake Peipus/Tchudskoe Ozero in 1917. This craft operated under the Estonian flag as *SA IX* after the end of the war (The photograph of this craft supplied by the author was not suitable for reproduction Ed.). The majority of these craft were seized by the Red Army; nine were found either at Kiev or in the Pripet Marshes in February 1919 and served with the newly formed Dnieper Flotilla in operations against the Polish Army and General Wrangel's White Army in the Southern Ukraine. Normally armed with a single mg, some were fitted with an old short barrel 37mm Hotchkiss which enabled them to harass conventional river steamers with impunity. Thus they were to be found not only on the Polish frontier in the early 'thirties but towards the end of the decade on the Amur in the Far East. Little known craft of an obscure riverine force, they provided the new Soviet Navy with the knowledge and experience that went into the design of the successful motor gunboats of World War II.

Units of the Soviet Amur Flotilla about 1939. Note BK5 with modified bow.

All Photographs from Dipl-Ing René Greger.

R-12 (SS-89)
APR 1942 off Key West, FL. National Archives

US SUBMARINE LOSSES

Vernon J Miller continues his analysis of US submarine losses in the Pacific

KETE (SS-369)

LCDR Edward Ackerman. All 87 crew lost.

Last Transmission Date: On 20 March 1945 at 29°, 38′N, 130°, 02′E after departing Guam 1 March 1945 for second patrol.

Loss Cause: Two loss possibilities exist; neither with sufficient evidence to permit any definite conclusions:
1. Possibly mined on or about 20 March 1945 south of Yakushima in minefield laid by Japanese minelayer *Tokiwa* and auxiliary minelayer *Koei Maru*.
2. Possibly torpedoed and sunk sometime between 20–23 March 1945 east of Okinawa by one of the four Japanese submarines lost in the area at about the same time.

Supporting Evidence:
1. Field of about 1000 mines was laid 27 February 1945. *Kete* was moving through the area at time when minefield was at about its most hazardous.
2. Japanese submarines *RO-41*, *RO-49*, and *RO-56* left for patrol in Okinawa area on 18 March and *I-8* left for patrol in same area on 20 March. All would have passed through area of *Kete*'s location. All were lost and none mentioned attacks against submarines in final reports.
  RO-41 (LT Yoshikuni Honda) is the Japanese submarine most likely to have encountered and sunk *Kete*. *RO-51* was sunk at 22°, 27′N, 132°, 19′′E on 23 March and had possibly been unable to report attack prior to her own loss. *Kete* had been ordered to leave area on 20 March and to proceed back to Midway. Could have been en route and surfaced when attacked.

LAGARTO (SS-371)

CDR Frank D Latta. All 85 crew lost.

Last Transmission Date: On 3 May 1945 after departing Subic Bay 12 April 1945 for second patrol.

Loss Cause: Depth charged and sunk 3 May 1945 in the Gulf of Siam, at 07°, 55′N, 102°, 00′E, by Japanese minelayer *Hatsutaka*.

Supporting Evidence: *Hawkbill* (SS-366) torpedoed and sank *Hatsutaka* 16 May 1945. Therefore, no Japanese records survive to support or detail this action. *Lagarto* was apparently caught in only 30 fathoms of water by an alert, aggressive, and well trained anti-submarine crew.

PERCH (SS-176)

LCDR David A Hurt. All 64 crew survived.

Last Transmission Date: On 27 February 1942 after departing Port Darwin 3 February 1942 for second patrol.

Loss Cause: Scuttled 3 March 1942 after being depth charged and heavily damaged 1 March 1942 north of Java, about 73 miles west of Bawean Island, Java Sea, at 113°, 50′E, 06°, 30′S, by Japanese destroyers *Amatsukaze* and *Hatsukaze*
  Detected while surfaced 2 March 1942 by Japanese destroyer *Ushio*. Thirty depth charges dropped between 0559hrs and 0830hrs, resulting in additional heavy damage.
  Detected again while surfaced 3 March 1942 by Japanese destroyer *Ushio* which opened main battery gunfire at 0659hrs, assisted by destroyer *Sazanami*.
  Then scuttled to prevent any possible enemy utilisation.

PICKEREL (SS-177)

LCDR Augustus H Alston Jr. All 74 crew lost.

Last Transmission Date: Never heard from after departing Midway 22 March 1943 for seventh patrol.

Loss Cause: Probably depth charged and sunk 3 April 1943 south of Hachinohe Sea, off Shiramuka Lighthouse, northern Honshu, by Japanese minelayer *Shirakami* and auxiliary subchaser *Bunzan Maru*, following bombing attack by Japanese naval aircraft.

Supporting Evidence: Japanese reported that aircraft detected oil patch at approximately 1700hrs and dropped two anti-submarine bombs at 1712hrs. *Shirakami* and *Bunzan Maru* dropped 26 depth charges. Japanese reported much oil surfaced. No other submarine was near area of attack.
  Pickerel had just sunk subchaser Ch-13 at 41°, 03′N, 141°, 58′E, and Japanese believe *Pickerel* may have been damaged by a mine following this sinking, thus explaining the initially detected oil.

Analysis: The Japanese cargo vessel *Fukuei Maru* was lost 7 April 1943 at 41°, 00′N, 142°, 00′E. If this ship was indeed sunk by a submarine and not lost accidentally as claimed by some Japanese sources, no other submarine could have been responsible. This would indicate *Pickerel* was not lost during the attack of 3 April. Although officially credited to *Pickerel* by JANAC, the actual loss cause remains doubtful.

POMPANO (SS-181)

LCDR Willis M Thomas. All 76 crew lost.

Last Transmission Date: Never heard from after departing Midway 20 August 1943 for seventh patrol.

Loss Cause: Two loss possibilities exist; neither with sufficient evidence to permit any definite conclusions:
1. Probably mined sometime between 29 August 1943 and 27 September 1943 off northern Honshu.
2. Possibly sunk 17 September 1943 by bomb and depth charge attack in Shiriyasaki Sea, off Aomori Prefecture, NE Honshu, by Japanese naval aircraft and minelayer *Ashizaki*.

Supporting Evidence:
1. Areas east of northern Honshu and between Honshu and Hokkaido were heavily mined. Loss to mine is considered the more probable cause.
2. Japanese claim that Ominato-based seaplane attacked surfaced submarine, which returned gunfire. Oil was seen to surface after attack. *Ashizaki* dropped depth charges the next day, bringing more oil to surface.

Analysis: Japanese cargo vessel Taiko Maru was lost 25 September in the Sea of Japan, west of Tsugaru Strait, at 41°, 30′N, 139°, 00′E. This sinking is credited to *Pompano*, which would indicate that no loss occurred in attack of 17 September. However, there is also distinct possibility that *Taiko Maru* was actually sunk by *Wahoo* (SS-238).

R-12 (SS-89)

LCDR Edward E Shelby. 42 Crew lost. Six saved.

Last Transmission Date: Transmission date non-applicable. Had not previously completed a war patrol.

Loss Cause: Sank within 15 seconds in 600 feet of water between 1220 and 1225hrs 12 June 1943 off Key West, Florida, at 24°, 24/30′N, 81°, 38′/30′W, probably from rapid flooding of forward part of ship through torpedo tube.
  Salvage not attempted due to depth of water.

ROBALO (SS-273)

LCDR Manning M Kimmel. 77 crew lost. Four saved.

Last Transmission Date: On 2 July 1944 after departing Fremantle 22 June 1944 for third patrol.

Loss Cause: Sunk 26 July 1944 two miles off western coast of Palawan, Balabac Strait, at 08° 25′N, 117°, 53′E, as result of what survivors described as explosion of after battery, but which more likely was a mine laid by Japanese minelayer *Tsugaru*.

Supporting Evidence: *Tsugaru* laid numerous mines in Balabac Strait area in March 1944 to strengthen original field laid in March 1943.

RUNNER (SS-275)

LCDR Joseph H Bourland. All 78 crew lost.

Last Transmission Date: Never heard from after departing Midway 28 May 1943 for third patrol.

Loss Cause: Several loss possibilities exist; none with sufficient evidence to permit any definite conclusions:
1. Probably mined sometimes between 8 June 1943 and 4 July 1943 off northern Honshu.
2. Possibly sunk 6 June 1943 by bomb and depth charge attack off Hakuto Lighthouse, in NE Honshu area, by Japanese naval aircraft and auxiliary submarine chaser *Minakami Maru*.
3. Possibly sunk 16 June 1943 by bomb and depth charge attack in Kamaishi Sea area by Japanese naval aircraft and destroyer *Nokaze*.
4. Possibly sunk 22–23 June 1943 by bomb and depth charge attack off Hakuto Lighthouse, in NE Honshu area, by Japanese naval aircraft, minelayer *Shirakami*, and auxiliary guard boats *Kaiwa Maru*, *Minakami Maru*, and *Miya Maru*.

Supporting Evidence:
1. Four known minefields were in assigned patrol area. Loss to a mine is considered the more probable cause.
2. Japanese claim considerable oil surfaced following this attack.
3. No further details mentioned in Japanese records. Apparently, action was incomplete and contact evidence was slight.
4. Apparently a concentrated attack over two-day period with 59 bombs and 66 depth charges being dropped. Japanese records claim oil surfaced as result of these attacks.

Robalo (SS-273)
15 Oct 1943, on trials in Lake Michigan. *Manitowoc Maritime Museum*

Pickerel (SS-177)
22 Dec 1942 at Mare Island, CA. *National Archives*

S-28 (SS-133)
18 Jun 1943 off Seattle WA. Final configuration. *National Archives*

Analysis: The Japanese passenger-cargo vessel, *Shinryu Maru*, was lost 26 June 1943 at 48°, 06′N, 153°, 15′E. If this ship was indeed sunk by a submarine and not lost accidentally as claimed by some Japanese sources, no other submarine could have been responsible. This would indicate *Runner* was not lost during any of the mentioned attacks, although loss after the sinking of *Shinryu Maru* is a definite possibility.

S-26 (SS-131)

LCDR Earle C Hawk. 46 crew lost. Three saved.

Last Transmission Date: Transmission date non-applicable. En route to patrol area for second patrol.

Loss Cause: Sank within few seconds after being rammed on starboard side at 2223hrs 24 January 1942 in Gulf of Panama, about 14 miles west of San Jose Light, by US subchaser *PC-460*. Attempts at salvage were unsuccessful due to 301-foot depth of water.

S-27 (SS-132)

LT Herbert L Jukes. All 49 crew saved.

Last Transmission Date: On 19 June 1942, reporting grounding, after departing Dutch Harbor 12 June 1942 for first patrol.

Loss Cause: Grounded at approximately 0043hrs 19 June 1942 on rocks off St Makarius Point, 400 yards off Amchitka Island, Aleutians.
  Abandoned by 1550hrs; the stranded hulk was left to be broken up by the pounding seas.

S-28 (SS-133)

LCDR Jack G Campbell. All 50 crew lost.

Last Transmission Date: Transmission date non-applicable. Had previously completed seven patrols.

Loss Cause: Operational loss from unknown cause shortly after 1820hrs 4 July 1944 off Lahaina, Honolulu, Hawaii, at 21°, 20′N, 158°, 23′W.
  Submerged at 1730hrs and sank in about 8400 feet of water while engaged in training exercises with Coast Guard cutter *Reliance*.
  No attempts made at salvage operations due to extreme depth of water.

S-36 (SS-141)

Lt John R McKnight Jr. All 47 crew saved.

Last Transmission Date: On 20 January 1942, reporting grounding, after departing Manila 30 December 1941 for second patrol.

Loss Cause: Grounded at 0404hrs 20 January 1942 on Taka Bakang Reef, in Makassar Strait, west of southern Celebes.
  Abandoned by 1330hrs 21 January 1942; rigged to flood to prevent any possible enemy utilisation.
  Complement rescued by Dutch launch *Attla* and the Dutch steamer *Siberoet*.

S-39 (SS-144)

LT Francis E Brown. All 44 crew saved.

Last Transmission Date: On 14 August 1942, reporting grounding, after departing Townsville, Australia, 10 August 1942 for fifth patrol.

Loss Cause: Grounded night of 13-14 August 1942 on submerged reef off Rossel Island, Louisiade Archipelago.
  Abandoned 16 August 1942; stranded hulk left to be broken up by the pounding sea.
  Entire complement rescued by HMAS *Katoomba*.

S-44 (SS-155)

LCDR Francis E Brown. 55 crew lost. Two survived.

Last Transmission Date: Never heard from after departing Attu, Aleutian Islands, 26 September 1943 for fifth patrol.

Loss Cause: Sunk by gunfire 7 October 1943 18.6 miles NNE of Araito Island, east of Kamchatka Peninsula, Kurile Islands, by Japanese patrol-escort vessel *Ishigaki* (LT Suekichi Seto).
  Survivors give the time of action as shortly after 2030hrs while the Japanese have recorded it as approximately 1832hrs.

Supporting Evidence: *Ishigaki* had departed Kashiwarable Bay, Shimushu Island, as escort for cargo ship *Koko Maru*.
  Lookouts sighted surfaced submarine which was taken under main battery and machine gun fire. First shot scored direct hit under conning tower. Range was closed to only 25 metres, and at least four other direct hits were scored.
  S-44 returned gunfire, but Japanese records indicate *Ishigaki* received only one machine gun bullet hit. At least 8 to 10 personnel were seen on deck, but only two were recovered by Japanese following the action.

Analysis: Survivors report that *S-44* had surfaced, under the mistaken impression that *Ishigaki* was a small merchant vessel and had opened fire on the target prior to any Japanese gunfire. However, it is equally possible that *Koko Maru* had been sighted originally and that presence of *Ishigaki* was not known.

***S-39* (SS-144)**
13 Jun 1930, underway of Tsingtao, China. Only wartime changes for this ship and *S-26* (SS-131), *S-27* (SS-132), *S-36* (SS-141) and *S-39* (SS-144) were painting out of white ID numbers and repainting entire ship black. Wartime photos of these units have not been located and probably do not exist.   *National Archives*

Ship Lists as Source Material (Locations)

Jan Glete puts forward a preliminary list of titles of value to all those interested in naval research

Historical ship-lists are valuable for studies of naval strength, shipbuilding activity, the navies as markets for naval stores, standardisation of warship design, the fluctuating demand for manpower, history of naval technology, etc. This bibliography has been prepared as a preliminary inventory of published sources for an international comparative study of the navies in the period 1500–1980, which will provide an overall view of the quantitative aspects of naval shipbuilding, naval strength and the demand for manpower, timber, mast trees, guns etc. For obvious reasons such a study must be based on ship-lists already prepared by others.

This bibliography only includes *historical* ship-lists, namely, lists which cover a number of years. In the numerous works devoted to naval history there are, of course, a large number of lists showing the state of a navy at a certain date and there are also many books and published papers which contain information from which historical ship-lists may be prepared. Such works are not considered here.

There are two main types of lists, alphabetical and chronological, the latter often arranged by ship types. Chronological lists are suitable for studies relating to the size and structure of a navy while an alphabetical list is useful as an index. Usually, they give some (seldom all) of the following particulars:

- name of vessel
- type and official classification
- building period (date of keel laying, launching or building year)
- building yard, shipwright or designer
- some notes about the vessel itself; renamings, battle honours etc
- final fate
- tonnage
- displacement (seldom)
- dimensions
- horsepower and speed (nineteenth-century steam vessels)
- number of crew
- armament (number of guns, established armament or armament actually carried on various dates)
- cost (very seldom given)
- source of information (rarely given)

It must be stressed that this paper in no way pretends to be complete. It is prepared as a working paper in order to invite comments and additional information from anyone who is interested in the subject. The bibliography is especially inadequate on Latin American and Mediterranean navies which reflects the fact that literature about them is difficult to find in northern Europe. The Far Eastern navies (China, Japan, Korea) are deliberately excluded. The same cannot be said regarding the navies of the Indian sub-continent. At least during the seventeenth century, European-style warships were built for Indian rulers but I am not aware of any ship-lists. Lists of warships and armed merchantmen belonging to the European East India Companies are included as far as I know of them but I have not made a systematic search for such lists.

The ship-lists in this article are published from the 1830s to the 1980s and there are considerable variations in scope and reliability. Some of them are definitive studies while others are only conjectural. They have been prepared by naval officers, naval historical departments, shiplovers and, to an increasing degree, academic historians. Lists which have been superseded by later publications are omitted for the sake of brevity.

ABBREVIATIONS
LMW   Lists of Men-of-War 1650–1700, parts 1–5, Society for Nautical Research London 1935–39
TBS   The Belgian Shiplover
FPDS   Foreign Periodical Data Service Newsletter, published by International Naval Research Organisation

ARGENTINA
This navy is well covered in a 7-volume work by Pablo E Arguindeguy, *Apuntes sobre los buques de la Armada Argentina (1810–1970)*, Buenos Aires 1972. A chronological list, with technical details and ship biographies, it also includes references to archival sources and literature. A short list is Christian de Saint Hubert, 'The Argentine Navy 1849–1974', TBS 1974, reprinted in Warship Supplement No 40. (Published by the World Ship Society.)

AUSTRALIA
A few colonial warships existed before 1860 and they are included in Ross Gillett, *Warships of Australia*, Adelaide 1977. *Australian Colonial Navies*, Colin Jones, Canberra 1986.

AUSTRIA
Chronological list: Karl Gogg, *Österreichs Kriegsmarine*

SHIP LISTS AS SOURCE MATERIAL (LOCATIONS)

1440-1848, Salzburg 1972; Karl Gogg, *Österreichs Kriegsmarine 1848-1918*, Salzburg 1967, revised edition 1974. These two books give details of dimensions, displacement (nineteenth century), armament, crew and speed (nineteenth century) as well as short ship biographies. Note that these books are intended to cover all the naval forces maintained by the Austrian Habsburgs.

One force which is not covered is the Imperial navy in the Baltic 1628-32. This is listed in *Sveriges krig 1611-1632*, Bilagsband I, Stockholm 1937. The list has been translated into German in part VIII of *Deutsche Militärgeschichte 1648-1939* (1977), p 18.

BELGIUM (Independent since 1830)
1830-1862: P Scarceriaux in TBS Jan/Feb 1954.

BRAZIL
The main work is by Lucas Alexandre Boiteux, *Das nossas naus de ontem aos submarinos de hoje (1822-1946)*, published in *Subsidios para a historia Maritima do Brasil*, vol XVIII to XXIV, Rio de Janeiro 1959 to 1971; concise ship biographies arranged alphabetically. There is also Christian de Saint Hubert's, 'Steam Warships built in Brazil 1822-1889', FPDS Newsletter No 2, 1980.

CHILE
Alphabetical list with brief ship histories, no data, Horace Via Valdivieso, *Manual de Historia Naval de Chile*, Valparaiso 1972.

Reportedly, there is a complete ship-list in Carlos Lopez Urrutia, *Historia de la Marina de Chile*, Santiago 1969, (I have not been able to consult it). Stephen S Roberts has prepared an unpublished list of powered warships, *The Chilean Navy (1847-1975)*: Gives building year, displacement, speed, armament and fate.

DENMARK (*Denmark-Norway* until 1814)
Before 1588: No list exists. For the period 1559-88 there is much information to be found in H D Lind, *Fra kong Frederik den andens Tid*, København 1902.

1588-1648: Alphabetical list: H D Lind, *Om Kong Christian den Fjerdes Orlogsflaade*, Tidsskrift for Søvaesen 1890. In later literature there is additional information but no attempt has been made to produce a revised list.

1650-1700: LMW Part 3, London 1936, by Preben Holck. Generally reliable with data and fate, but it is useful to consult Jörgen H P Barford, *Orlogsflåden på Niels Juels tid 1548-1699* (1963). There is no ship-list in the strict sense of the word in Barford's book, but all vessels (including hired merchantmen and small yard-craft excluded by LMW) are mentioned in the text.

1700-1830: Chronological lists in H G Garde, *Efterretninger om den danske og norske sömagt I-IV*, København 1832. Date of launching, designer, dimensions, number of guns, crew, final fate.

From 1807 onwards: Lists published by H Degenkolv (under various titles) in 1867, 1889 (with addendas 1893, 1899 and 1906). With data and fate.

1814-1848: J H Schultz, *Den danske marine 1814-1848 I-II* (København 1930-32). A detailed list as appendix to a detailed technical history. The construction of warships at the Copenhagen naval dockyard (where most warships were built) is listed in *Tidskrift for Søvaesen* 1923 and in *Skibsbygning og Maskinvaesen ved Orlogsvaerftet... 1692-1942*, København 1942. These lists start in 1692 and give date of keel-laying, launching, number of guns, dimensions and (only 1923 list) displacement and final fates.
Birget E Thomsen, 'Danish Oared Gunboats 1700-1850', *Matinehistorisk Tidskrift No 4, 1977*
Stephen S Roberts, 'Steam Warships of the Royal Danish Navy' was published in TBS no 133 (1970): building year, displacement, speed, guns, final fate.

ENGLAND/GREAT BRITAIN
Alphabetical list: J J Colledge, *Ships of the Royal Navy. An Historical Index, Volume 1 (Major ships) & Volume 2* (1969-70). Date of launching, tonnage/displacement, dimensions, builder, number of guns, final fates. Another alphabetical list is T D Manning & C F Walker, *British Warship Names* London 1959. Less complete than Colledge but interesting as a history of the origin of shipnames. J J Colledge published a supplement to his two-volume work under the auspices of World Ship Society 1986.

Early fifteenth century: Susan Rose, *The Navy of the Lancastrian Kings* (Publications of the Navy Records Society, vol 123, 1982).

1509-1649: R C Anderson, *List of English Men-of-War 1509-1649* (The Society for Nautical Research 1959). For the period 1539-1588 this list has been revised by Tom Glasgow in Mariner's Mirror No 3, 1970. It gives building year, yard, tonnage, dimensions, number of guns, fate.

1649-1702: LMW Part 1 by R C Anderson, same scope as 1509-1649. Somewhat more detailed lists are included in J R Tanner, *A Descriptive Catalogue of the Naval MSS* in the Pepysian Library, vol 1 (Publications of the Navy Records Society, vol 26, 1903). Period 1660-1688, includes depth in hold, draught, designer and manning establishment. Frank Fox, *Great Ships. The Battlefleet of King Charles II*, London 1980.

1702-c.1850 (excluding steamers): Fred Dittmar in TBS No 129 (1968) covers hired vessels 1793-1815, Nos 130-135 (1969) all warships 1702-1792, No 4, 1970 to No 2, 1972 all warships 1793-1850. Scope about the same as LMW but with more detail about armament but no dimensions given. From 1793 onwards year of ordering is listed.

There is no comprehensive list of British steam warships up to 1860. Steam battleships are covered in detailed lists in Andrew Lambert, *Battleships in Transition. The Creation of the Steam Battlefleet 1815-1860*, London 1984. Steam gunboats from 1854 onwards, Antony Preston & John Major, *Send a Gunboat*, London 1967.

Sailing ships-of-the-line 1650-1850: a detailed list in Brian Lavery, *The Ship of the line*, vol 1, London 1983. Auxiliaries are listed in EFS Fisher, *List of Admiralty Storeships, Storecarriers, Store Transports & Fuelling Vessels 1600-1962* (1962) and in E E Sigwart, *Royal Fleet Auxiliary* (1969).

There are also some lists which cover warship building at a particular naval dockyard: James Goss, *Portsmouth-built warships 1497-1967* (1986) and K V Burns, *Plymouth's Ships* of War: A History of Naval Vessels Built in Plymouth between 1694 & 1860 (1986).

Warship losses, chronological list: G R Ransome, *British Naval Ship Losses from 1700*, in Warship Supplement (published by World Ship Society) Nos 1-30.

For the Scottish Navy which existed in the sixteenth century (before the Union with England) there are apparently no lists published. (There is at least one published work. *The Old Scots Navy*, Ed.)

Ships of the British East India Company are listed in Jean Sutton, *Lords of the East. The East India Company and its Ships*, London 1981. The list includes tonnage, life-span and number of voyages. The Source on the Indian Navy is Lt C R Low, History of the Indian Navy, (London 1877). See also *Australia* and *Indian Ocean Area*.

France
Alphabetical list: *Répertoire des Navires de Guerre Francais* (1967), edited by Jacques Vichot (based on research by Pierre Le Conte and Jean Meirat). Gives building period, building yard, type, final fates and battle honours. This book also lists privateers, East Indiamen and merchantmen involved in naval actions. The only drawback is that the distinction between private-owned and state-owned vessels is not always made clear. Some privately owned vessels are listed as if they were part of the navy.

Chronological lists: 1648—1700: LMW, Part 2 by Pierre Le Conte.

1774-1783: Ships-of-the-line and frigates listed in an appendix to Jonathan R Dull, *The French Navy and American Independence* (Princeton 1975). *1814-1850*: Sailing warships (except the smallest) listed by Stephen S Roberts in TBS no 156. The list gives building period, building yard, dimensions, displacement, armament and fate. Cdt de Balincourt and Pierre Le Conte prepared lists of early French steam warships which were published in *Revue Maritime*: paddle-wheel steamers are listed in nos 10/1932 and 1/1933. Steam ships-of-the-line in nos 3 & 5, 1933. These lists give building dates, displacement, dimensions, armament, crew, horsepower, speed, designer, builder, machinery manufacturer, operational history and fate.

Steam ships-of-the-line are listed in Andrew Lambert, *Battleships in Transition*, London 1984, whilst screw-driven cruisers are listed in Christian de Saint Hubert, *Builders, Engine Builders and Designers of French Screw Frigates, Corvettes and Despatch Vessels (1844-1887)*, FPDS Newsletter no 3, 1984. This list also gives displacement and horsepower. Mr de Saint Hubert has prepared a similar list for French paddle-wheel warships, still unpublished.

Ships in the French naval shipyards are listed in Paul Coat, *Les arsenaux de la Marine de 1630 à nos jours*, Brest-Paris 1982. It is not complete, however, as warships built in Lorient before 1861 are omitted (see also literature on the French East India Company below). The naval shipbuilding effort in French Canada is not included, but it is covered in depth by Jacques Mathieu, *La Construction navale royale a Québec 1739-1759*, Quebec 1971, including a ship-list appendix.

Jean Boudriot has published several lists of various types of warships, usually with details about dimensions, designer, yard, guns and final fates: Three-decked First Rates: *Neptunia* 102. 80-gun ships: *Neptunia* 151. 64-gun ships: *Neptunia* 142. Frigates with 18-pounders: *La Venus 1782*. Frigates with 12 pounders: *La Belle Poule 1765*. Barques longues: *Neptunia* 117. Brigs: *Le Cygne 1806-1808*. Brigs: *Le Cygne 1806-1808*. Bomb vessels: *Neptunia* 99 and in *Le Salamandre*. Establishments of Ordnance 1674-1849: *Neptunia* 103.

Jean Boudriot has also dealt with the French East India Company's ships in *Compagnie des Indes 1720-1770. Vaisseaux, hommes, voyages, commerces*, Paris 1983, where a detailed list of ships built for the company at Lorient is included. A less detailed but more comprehensive list of the company's ships 1717-70 (including purchased ships) has been prepared by Geneviéve Beauchesne and published in *Inventaire des archives de la Compagnie des Indes*, Paris 1978.

Germany (except Austria)
There were several cities and states in Germany which maintained navies during the period 1500-1860 but after the early sixteenth century there were no major naval powers in this area. During the eighteenth century and the first half of the nineteenth there was no force at sea which could be termed a standing navy. The following historical ship-lists are known to me:

Lübeck 1563-70 (during the war with Sweden): Herbert Kloth, *Lübecks Seekriegswesen in der Zeit des nordischen siebenjähriges Krieges 1563-1570*. Zeitschrift des Vereins für Lübeckische Geschichte und Altertumskunde 1923. A detailed description of the ships.

Brandenburg, Kurland, Hamburg, Bremen, Lübeck 1643-1700: LMW, part 3 by W Vogel & H Szymanski. Building year and place, dimensions, guns, final fate. Hans Szymanski, *Brandenburg-Preussen zur see 1605-1815*, Leipzig 1939, comprehensive. 1815 onwards: Erich Gröner, *Die Deutschen Kriegschiffe 1815-1945*, München 1966, revised edition 1983, G Kroschel & A-L Ewers, *Die Deutsche Flotte 1848-1945*, Wilhelmshaven 1962 and later editions, Walter Hubatsch, *Die Erste Deutsche Flotte 1848-1853*, Herford 1981, and Gerd Stolz, *Die Schleswig-Holsteinische Marine 1848-1852* (1978). Detailed descriptions of the ships in most cases.

There is also a seven-volume work by Hans H Hildebrand, Albert Röhr & Hans-Otto Steinmetz, *Die Deutschen Kriegsschiffe. Biographien – ein Spiegel der Marinegeschichte von 1815 bis zur Gegenwart*, Herford 1979-1983, where detailed ship histories are arranged alphabetically.

Greece
The main work is an alphabetical list by K Paizis-Paradelis, *Ta ploia tou ellenikou polemikou nautikou 1830-1979* (*The Ships of the Greek War Navy 1830-1979*), 1979. This gives a short biography of every ship with a few technical details.

Stephen S Roberts has prepared an unpublished paper, *Steam Warships of the Greek Navy* (1980) which gives building year, displacement, speed, armament and fate.

Indian Ocean Area
In British India a local Royal Indian Marine was organized. The warships of this navy may be found in Colledge's alphabetical index of the Royal Navy. A list of steamships from 1832 onwards prepared by Ian Grant is published in Naval Notebook No 4, 1979. Of the navies belonging to non-European rulers an initial effort has been made in respect of the Oman-Muscat-Zanzibar navy in *Oman - A Seafaring Nation* (1979). This book gives fragmentary lists from the seventeenth to nineteenth century.

SHIP LISTS AS SOURCE MATERIAL (LOCATIONS)

ITALY (except Venice)
In the sixteenth and seventeenth centuries most of Italy belonged to the Spanish Habsburg empire and the navies maintained here are normally regarded as parts of the Spanish navy. For the eighteenth and nineteenth centuries (up to 1860) Lamberto Radogna, *Cronistoria delle Unita da Guerra delle marine preunitarie*, Rome 1981, covers Naples and the Two Sicilies 1733-1860, Sardinia 1815-1860, Toscana 1737-1860 and the Papal State 1775-1870. Also includes the revolutionary navies of Sicily in 1848-49 and 1860, Plus the Sardinian lake flotilla of 1859-81. Alphabetical lists with detailed ship histories, data, when known.

MALTA
From 1530 to 1798 the island was the home of the Order of St John of Jerusalem which maintained a navy to fight Barbary pirates, Muslim warships and privateers. This navy has recently been dealt with in Ubaldino Mori Ubaldini, *La Marina del Sovrano Militare Ordine di San Giovanni di Gerusalemme di Rodi e di Malta*, Rome 1971. There is no list of warships, but a list of ship's captains and the ships to which they were appointed, gives an approximate view of the navy's size and structure.

MEXICO
Steamers are listed in Christian de Saint Hubert, *Notes on the Mexican Steam Navy (1842-1945)* published in Warship Supplement No 57 (World Ship Society).

NETHERLANDS
There are no historical ship-lists covering the large navy maintained by the Dutch Republic before 1648. The 1648-1702 period is treated in LMW part 4, by A Vreugdenhil: building year, admiralty, dimensions, number of guns, fate. Warships of the Amsterdam Admiralty 1713-1751 are listed in J R Bruijn, *De Admiraliteit van Amsterdam in Rustige Jaren 1713-1751* Amsterdam & Haarlem 1970: building year, master shipwright, number of guns, dimensions and final fate. From 1814 onwards A J Vermeulen, *De Schepen van de Koninklijke Marine en die der gouvernementsmarne 1814-1962* (1962) gives a practically complete list although with few details about the sailing ships: building year, building yards, number of guns, a short biography.
  C W J Schaap, *Warships of the Friesland Admiralty in the 17th and 18th Centuries*. 1982 Yearbook, Fries Scheepvaart Museum en Oudheidkamet.

NORTH AFRICA
The navies of this area - Algeria, Egypt, Morocco, Tunis and Tripoli - were practically independent forces, although the countries nominally belonged to the Turkish Empire. As the Barbary coast vessels were a considerable nuisance to European and American merchantmen these navies have attracted considerable interest, but in the literature I have been unable to locate anything which may be called a ship-list.

NORWAY
The book *Norges Sjöforsvar 1814-1914* (1914) does not have any ship-list but all major warships of the period are described. In *Skipsbygning på Horten gjennom 150 år 1818-1968* (1968) there is a list of ships built at the Navy's main dockyard. In TBS No 155 Stephen S Roberts has published a list of powered vessels 1840-1975: building year, displacement, speed, armament and fate.

PERU
Steamers are listed in Stephen S Roberts, *The Peruvian Navy (1847-1977)*, an unpublished list: building year, displacement, speed, armament and fate.

POLAND
This country maintained a navy periodically during the first half of the seventeenth century. A list of the navy during the 1620s is to be found in *Sveriges krig 1611-1632, Bilagsband I*, Stockholm 1937, and in Eugeniusz Koczorowski, *Bitwa pod Oliwa* Gdansk 1976. The navy of the 1630s is listed in Wladyslaw Czaplinski, *Polska a Baltyk w Latach 1632-1648. Dzieje floty i polityki morskiej*, Wroclaw 1952.

PORTUGAL
The most comprehensive list is António Marques Esparteiro, *Catálogo dos Navios Brigantinos (1640-1910)*, Lisbon 1976, which lists building year, building place, number of guns and fate of all Portuguese warships 1640-1910. The same author has also published numerous ship histories in the Series *Colecção Estudos* published by the Portuguese Ministry of Marine. They are titled *Três séculos no mar*, Lisbon 1974-1986. The following parts cover warships built up to 1860:
  Vol 3: Caravelles and Galleons (seventeenth century)
  Vol 4-9: Ships-of-the-line
  Vol 10-13: Frigates
  Vol 14-17: Corvettes
  Vol 18-19: Brigantines and brigs
  Vol 20: Transports
  Vol 22: Schooners and yachts
  Vol 23: Small, early steamers
  Vol 24: Steam gunboats
  In the English language J O Ramos published a list in TBS 84 and 88. It has much the same scope as *Catalogo* although it represents an earlier stage of research.
  There are no historical ship-lists extant for the sixteenth century or for the period 1580-1640 when Portugal was a part of the Spanish Habsburg empire.

RUSSIA
Feodor F Veselago, *Spisok Russkich Voennich Sudov 1668-1860* (1872). A complete list of Russian warships until 1860 with dates of keel-laying and launching, building yard, designer, dimensions, armament and fate. Parts of the list have been translated into German in Andreas Bode, *Die Flottenpolitik Katharinas II und die Konflikte mit Schweden und der Türkei 1768-1792*, Wiesbaden 1979. Steam ships-of-the-line are listed in Andrew Lambert, *Battleships in Transition*, London 1984.

SPAIN
For the period before 1700 there are no historical ship-lists covering the decentralised navies of the Spanish Habsburg empire. Pierre and Huguette Chaunu, *Seville et l'Atlantique 1504-1650*, Paris 1955-1960, gives annual lists of ships serving as escorts to the convoys sailing to and from America. For the period after 1700 there is a list of ships-of-the-line in Gervasio de Artinano y de Galdacano, *La Arquitectura naval espanola (en madera)* (1920). The list gives building year and yard, number of guns and final fates. The early steamships are described in detail by Christian de Saint Hubert in *Warship International* No 4/1983 and No 1 1984. Mr de Saint Hubert has also prepared lists of Spanish ships-of-the-line 1714-1825 which are published in *Warship* Nos 37, 38 and 39. They show that the list in Artinano is in need of a certain revision.
(There are also lists of Spanish warships post 1700 in *El Buque en la Armada Espanola* (1981) but they are very unreliable).

SWEDEN
1521-1560: Jan Glete *Svenska örlogsfartyg 1521-1560* in Forum Navale No 30-31 (1976-77).
1611-1632: *Sveriges krig 1611-1632, bilagsband I*, Stockholm 1937
1645-1648: *Från Femern och Jankow till Westfaliska freden*, Stockholm 1948
1634-1680: Axel L Zettersten, *Svenska flottans historia II*, Stockholm 1903, a list with numerous errors.
1650-1699: LMW, part 3 Hjalmar Börjeson.
1680-1814, ships-of-the-line: Gunnar Unger, *Svensk sjökrigshistoria II*, Stockholm 1923, somewhat unreliable and with no technical details.
1700-1721: Lars O Berg in *Forum Navale* No 25 (1970). Lists the main fleet based in Karlskrona.
1756-1791: archipelago fleet: Oscar Nikula, *Svenska skärgårdsflottan 1756-1791*, Helsinki 1933, lists only named vessels, numerous small craft with number omitted.
1771-1814: Ships-of-the-line and frigates added during this period are listed in *Svenska flottans historia II*, Malmö 1943.
1808-1849, oared warships; Lars O Berg in *Forum Navale* No 24 (1968).
1850-1900: Lars O Berg in *Forum Navale* No 21 (1965).
Powered vessels are listed by Stephen S Roberts in TBS No 4, 1975. Vessels built on the western coast of Sweden 1700-1750 are listed in Ernst Bergman, *Gamila Varvet vid Goteborg 1660-1825*, Gothenburg 1954. Most of these lists give building place, building year, armament or at least number of guns, dimensions and fate.
The Swedish East India Company, (actually a company trading with China), existed from the 1730s to 1808. Its ships are listed in Hugo Hammar, *Fartygstyperna i Svenska Ost-Indiska Compagniets flotta*, Gothenburg 1931.

TURKEY
In spite of the fact that this country maintained a large navy for most of the period 1500-1860 there are no Turkish historical ship-lists available. Three steam ships-of-the-line are listed in Andrew Lambert, *Battleships in Transition*, London 1984, and Stephen S Roberts has prepared an unpublished and conjectural list, *The Turkish Steam Navy*.

USA
Alphabetical list: *Dictionary of American Naval Fighting Ships, vol 1-8*, Washington 1959-1981. Extensive ship biographies with tonnage dimensions, complement, armament and speed.
Chronological list: K Jack Bauer, *Ships of the Navy 1775-1669, Volume I: Combat Vessels* (1970) (no further volume has appeared). This list gives date of keel-laying, launching and commissioning, builder, engine manufacturer, fate, tonnage or displacement, dimensions, armament, horsepower, speed and complement. Although no ship-list is included, Howard I Chapelle, *The History of the American Sailing Navy* (1949) gives detailed descriptions of all sailing warships.
Register of Officer Personnel and Ships Data 1801-1804, Washington 1945. Data on the ships of the USN during the years stated. Bennett Frank, *The Steam Navy of the United States*, Pittsburg 1896. 40th Congress. 2nd Session 7th May 1868. *Table of Naval Vessels as at 1st April 1861. Table of Naval Vessels built for the Navy Dept since 1st April. Table of Naval Vessels purchased for the Navy Dept since 1st April 1861*, Washington 1868.

VENICE
Major sailing warships are listed in Cesare Augusto Levi, *Navi da Guerra construite nell'arsenale di Venezia dal 1664 al 1896*, Venice 1896. There are some more or less obvious minor errors apparent in this list, possibly due to misprints. It is condensed and translated into English by Ilo Barbensi in TBS no 3, 1974. These lists give date of launching, length of keel, rate and fate.

This short summary shows that only the British navy is covered by chronological lists for most of the period 1500-1860. Russia, USA, Italy, Austria and several new navies of the nineteenth century; Argentina, Brazil, Chile, Greece and Prussia are well covered by chronological and alphabetical lists. The Danish navy is relatively well covered from the end of the sixteenth century and Portugal from 1640. France, Sweden and Venice are only partially covered by chronological lists while major seapowers such as Spain and the Netherlands are to a large extent unrepresented. The navies of Turkey, Egypt and the Barbary Coast powers are little known. There are virtually no historical ship-lists for the large galley-fleets which operated in the Mediterranean in the sixteenth and seventeenth centuries. However, several major studies on galley warfare and galley technology make it possible to gain an approximate view of their size. Looking at the subject from a chronological viewpoint we find that the sixteenth century is represented mainly by England and Sweden, the seventeenth century by England, Sweden, Denmark and Portugal from 1640 plus Netherlands and France from 1648. Venetian ships-of-the-line are recorded from the 1660s onwards. From the eighteenth century we have a relatively complete picture, although there are large gaps in the chronological lists for Spain, France, Sweden (oared warships), the Netherlands and Venice (minor warships). The nineteenth century is fairly comprehensively covered although with gaps for Great Britain (steam cruisers), France (minor steamers), and Spain (sailing cruisers). The Muslim world remains something of a mystery even in more recent times.

The author wishes to thank Mr Christian de Saint Hubert, Nairobi, Kenya; Mr Adnrzej v Mach, Gdansk, Poland; Mr R H C van Maanen, The Hague, Netherlands, and Mr John P F H Cook (Mainmast Books) for very valuable help in the preparation of this bibliography.

Readers are invited to suggest additional sources for inclusion in this project.

A's & A's

Eagle Boats

DK Brown RCNC has unearthed a technical assessment of these craft prepared by Stanley Goodall RCNC, Department of Naval Architecture, Washington. 'On 26 December 1917 the Chief Constructor was directed to prepare a design for an anti-submarine patrol boat with the following military characteristics:
Armament: two 4in low angle guns.
One 3in AA gun.
One depth charge thrower & depth charge rails.
Speed:
18 knots
Radius of Action: 3500 nautical miles at 10 knots
Special qualities: listening equipment and ability to manoeuvre quickly.

Delivery was desired at a very early date, and from the first it was known that the building of these boats would be entrusted to a firm with no experience whatever of shipbuilding. These two factors affected the design very considerably. Simplicity, making for easy and rapid work, was sought in every detail; scantlings were fixed with a view to allowing a margin of strength to cover possible bad workmanship rather than to reducing the hull weight to a minimum.

The form was devised so that the waterlines were absolutely straight for a considerable length in the forward and after bodies, thus maintaining a constant bevel for the frame angles; the sides were straight, the rise of floor in the forebody and both the frame lines were also straight. Only one strake of plating, that at the turn of the bilge, required bending; and the straight frames at the side and bottom were bracketed together to avoid angle-smith's work. The deck beams had no round up and the sheer was provided by two straight lines so that the deck erections could be built as square houses, brought to the ship complete, and fastened down immediately.

The number of different sections and plate thicknesses were kept to a minimum. Stresses accepted in the Hogging condition (displacement 600 tons) on the assumption that the bending moment would be $\frac{W \times L}{25}$ were 4.7 tons per sq in tension and 3.9 tons per sq in compression.

The 4in guns were placed in high commanding positions and, in order that the ship might be a steady gun platform, a moderately low metacentric height of one foot, and a range of stability of about 70 degrees (allowing for free surface of oil and water) were accepted for the normal load condition. The rudder was given a large area to provide good manoeuvring qualities.

As the efficiency of the listening apparatus depended on the noiselessness of the main and auxiliary machinery, this received special attention. Steel wire hawsers were substituted for chain cables, and storage batteries were provided so that the vessel could dispense with the electric light generator for short periods.

Particulars of the design are as follows:
Length between perpendiculars: 200ft, breadth 25ft, 6in, draught 7ft, 3in.
Freeboard forward: 11ft, 3in, amidships 11ft, 3in, aft 7ft, 3in.
Displacement: 500 tons. SHP at 475 RPM = 2000. Fuel carried on load draught 60 tons. Fuel required to meet desired radius of action 90 tons. Total fuel carried 148 tons.
Weight of hull and hull fittings: 237 tons. Main machinery 82 tons. Armament 10 tons*.
Reserve feed water: 10 tons. Equipment, including electrics, anchors, cables, rigging etc 17 tons. Outfit (Boats, furniture and stores) 27 tons. After model experiments had been made in the tank it was decided that the flare of the forward sections could be eliminated, without making the boat unduly wet. At first it was hoped that bilge keels would not be required but they were added as a result of the rolling experienced in the Model Basin.

Design work commenced 26th December 1917. Preliminary drawings to builders – December 31st. Design completed 8th January 1918. On trials the first boat of the class satisfactorily fulfilled the requirements of the design.'

Note: The original assessment seems to have been compiled in haste. Ed.

The Loss of Bremse

Herr Werner F G Stehr, West Germany, responding to the article in *Warship* 41, has furnished the following information, based on material from official sources and the private archives of MAN, the marine engine builders.

Bremse was not fitted with 50 per cent of the diesel installation intended for the three panzerschiffe of the *Deutschland* class. The misunderstanding has arisen because her main motors

*This figure seems to be rather low? Ed.

produced half the shaft horsepower of the sets fitted to the larger vessels. She was launched in January 1931 and first commissioned for service on 14 July 1932, whereas *Deutschland* was launched in May 1931 and first commissioned on 1 April 1933. The motors for *Deutschland* were ordered in October 1928 whilst those for *Bremse* were ordered on 9 January 1930. They were of different powers as follows:
Bremse – eight two-stroke double acting MAN diesels, type M8Z 30/44, four two-stroke double acting MAN diesels type M4Z 30/44, for auxiliary functions.
Deutschland – eight two-stroke double acting MAN diesels type M9Z 42/58, four two-stroke double acting MAN diesels type M5Z 42/58 for auxiliary functions.
Type description reads 'Marinemotor, eight cylinders in line, two stroke, bore 30cm, stroke 44cm.' (zweitakt = two stroke)
Power output: Bremse. Eight sets at 3550bhp ea. Total 28400bhp or 26000shp = 29 knots for one hour/27 knots for a longer period. 8kg/bhp.
Deutschland. Eight sets at 7100bhp ea. Total 56800bhp or 54000shp = 28 knots for one hour/25 knots for a longer period. 11.5kg/bhp. The motors in *Bremse* were of the same type (with an additional cylinder) to those fitted to *Leipzig* for cruising purposes: four two-stroke double acting MAN diesels, M7Z 30/44, producing 3100bhp ea.

To add to the confusion, when first planned *Ersatz Drache* was intended to act as a test vehicle for the main motors of the panzerschiffe, but by the time her machinery was ordered from MAN the idea had been dropped and she was completed as a Gunnery Schoolship/Minelayer. She also acted as a training ship for diesel mechanics. Initially, she was conceived as a long range scout with a range of 8000 nautical miles at 19 knots. For this role a seaplane was to have been carried and two torpedo tubes. The wide spacing of the funnels was due more to the aircraft handling arrangements of the original design than the layout of the motor rooms, and was retained to give 'B' gun as wide an arc of bearing as possible. The 12.7cm outfit also stemmed from the original concept and was intended to give her a margin of superiority over hostile destroyers, a class of vessel denied the German Navy under the terms of the Treaty of Versailles. In the event, as first commissioned, she turned out somewhat unsteady as a gun platform and was immediately taken in hand for modifications to reduce top weight. The funnel casings were redesigned, not removed as stated in the article. That she was only capable of 23 knots in early 1940 was due to her main motors being well overdue for periodic overhaul as they had run over twice the number of hours laid down. *Carl Peters*, commissioned on 6 January 1940, was a modern, well-equipped depot ship for the 3rd Schnelle Torpedo-Motorboote Flotille with a top speed of 23 knots.

The motor installation intended for *Ersatz Drache* as first planned was to consist of four two-stroke double acting MAN diesels type M6Z 42/58 of 3100bhp, two to each shaft. Note: All the MAN diesels mentioned were of the reversing pattern when fitted as main engines.

Iowa Class

Mr Norman Friedman of New York has expressed concern at the content of the editorial in *Warship* 43 which he regarded as a somewhat bizarre interpretation of US policy. He also felt that the reasoning behind the reactivation of the *Iowa* class had likewise been misunderstood. As he is a US citizen his concern is understandable; however, as *Warship* is not a political forum in any sense of the word, interested parties are recommended to re-read his most interesting article 'The Reborn Battleships' in *Warship* 31, July 1984, and his very worthwhile *The Postwar Naval Revolution* (both available from Conway Maritime Press) and draw their own conclusions.

It would be appreciated if contributors to this feature would conform to the usual practice and submit their comments either in double spaced typescript or in clear handwriting with a broad left hand margin. Criticism of an author should be avoided. The editorial team select the content of *Warship* and must accept responsibility if factual errors find their way into print.

The Old 'S' Class Destroyers 1939–45

Commander H St A Malleson RN (Ret) of East Sussex has, on the basis of commissions in *Seraph, Vivacious, Campbell* and a spell in *Keppel* plus sea time in *Minion, Sturgeon* and *Sardonyx* (as Signal School Tender), forwarded the following comments on the 'S' Class (*Warship* 37):

The class lacked seakeeping qualities, a serious drawback in destroyers intended to work with the fleet.

The 'V' and 'W' Classes were built because the Admiralty believed, wrongly as it turned out, that the Germans were planning to build larger destroyers. The difference in cost compared with the 'S' Class was slight.

The 'S' Class were not much good on service in World War II mainly because of their low height of eye from the bridge – a grave disadvantage. The photo on page 21 shows the absurd structure fitted aft to carry an effective Radar 271 – height of 'eye' again.

Mr R Nailer of Dorset has also forwarded some interesting material on the 'S' Class, too long to quote in As and As but which will be published in *Warship* 46. It deals with the individual vessels and their record of service from about 1938 until the end of World War II.

BOOK REVIEWS

RUNNING TO THE SHROUDS – RUSSIAN SEA STORIES

Konstantin Stanyukovich

Translated by Neil Parsons
Forest Books, 215 × 140mm, 100 pp

ISBN 0 948259 06 X
£5.95 Paperback

This is one of those books where the introduction is of more interest than the content of the stories. The writer, a well known literary figure inside Russia, the son of a Vice-Admiral, former Naval Cadet and junior officer of barely four years experience in the mid 1860s, wrote the stories during a short period of Siberian exile in the 1880s. They paint a somewhat rosy view of life in the Russian Navy some 20 years earlier: noble band of brothers with an occasional bad Captain thrown in. Published in the 1890s, they were received with enthusiasm by the upper and middle classes and the reviewer is strongly reminded of the accounts of life at sea to be found in the *Strand Magazine* and similar journals of the same period. Suitable for the ladies, they contain none of the violence and brutality found in Marryat. Interesting because of the background.

BRITISH WARSHIPS & AUXILIARIES

1987/88 Edition
Mike Critchley

Maritime Books 210 × 150mm
112 pp, illustrated throughout

ISBN 0 907771 30 0
£3.95 Paperback

This is a fairly successful attempt to cram a quart into a pint pot. Some of the photographs are excellent, others less so. The editorial is of interest. The author makes the point that as ever, the British Armed Forces are penny wise and pound foolish when it is a question of retaining trained and experienced personnel. One senses the dead hand of the Treasury in the background. A good buy if you want a handy record of the Royal Navy.

THE ROYAL NAVY IN FOCUS – 1960-69

Editor **Mike Critchley**

Maritime Books 180 × 250mm, 160 pp

ISBN 0 907771 44 5
£4.95 Paperback.

An interesting selection of photographs from the collection of Messrs Wright and Logan. The page size favours the smaller craft and there are some splendidly sharp views of minor warships and support vessels. Value for money if you consider the price of individual prints today.

NAVAL CANNON
John Munday

Shire Publications, 210 × 150mm, 32pp, illustrated

ISBN 0 85263 844 2

£1.25 Paperback

One of a series of promising little books that provide good introductions to the subject of the titles, written by people with a sound grasp of the particular trade or industry. There are some well known names amongst them. The reviewer has seen a number of the titles and thoroughly commends them to the discriminating purchaser. Ideal for the young person with an enquiring mind.

JAPANESE NAVAL VESSELS OF WORLD WAR TWO

As Seen by US Naval Intelligence
Introduction by **A D Baker III**

Arms and Armour Press, 170 × 260 mm, well illustrated

ISBN 0 85368 847 8   £14:95

This is a most interesting book being a reprint of an essential aid to the US fighting seaman and naval airman actively engaged in the Pacific War, and it is a sad fact that ignorance of the content could lead to a sudden departure from this life. It contains in a concise format virtually everything the US Navy knew about the Japanese enemy outside the covert signals interception decoding effort. That much of the technical data is inaccurate is immaterial. It was vital that anything Japanese be identified at first sighting, whether it be in bright moonlight or swirling mist. Fortunately for the allies, the Japanese Navy possessed a collection of some of the most distinctive warships ever seen, as is made abundantly clear in these pages. Just browsing sent a cold shiver down the spine of the reviewer, who well remembers, as a young schoolboy, the attack on Pearl Harbour, the loss of *Prince of Wales* and *Repulse*, the fall of Singapore, the Battle of the Java Sea and the seemingly irresistible advance of the Japanese. A very useful book to have around when studying the war in the Pacific and the Far East, and a timely reminder of the threat posed by a ruthless and resolute enemy.

USCG ALWAYS READY
Hans Halberstadt

Blandford Publishing Ltd, 212 × 204mm 130 pp, well illustrated

ISBN 0 85638 867 2

£9.95 Paperback

An interesting but somewhat sensational account of that fine service, the United States Coastguard. The author is a documentary film maker and an accomplished photographer, but unfortunately his style can be irritating. Within its limitations the book does do justice to the service and its surprisingly wide range of responsibilities. It appears that although reasonably well funded the Coastguard suffers from not being able to control how its annual budget is spent - this rests with the Appropriations Committee of the Congress of the United States, which body will approve the financing of certain aspects of the service with enthusiasm, whilst cutting the finance for others to the bone. Only to change the priorities the following year! Inevitably, this leads to a lack of continuity and lowering of morale. Despite this the service maintains its high standards. So much so that many 'old hands' consider that it is its own worst enemy. Recommended.

Mainmast Books

Saxmundham, Suffolk, IP17 1HZ, England

Apart from stocking thousands of U.K. titles, we import nautical books from:-

Argentina
Australia
Austria
Belgium
Brazil
Canada
Chile
Denmark
Finland
France
D.R. Germany
F.R. Germany
Greece
Hong Kong
Hungary
Iceland
India
Rep. of Ireland
Italy
Japan
Kenya
Malaysia
Mauritius
Netherlands
New Zealand
Norway
Oman
Poland
Portugal
South Africa
Singapore
Spain
Sweden
Switzerland
U.S.A.

New, Secondhand and Antiquarian Marine Books

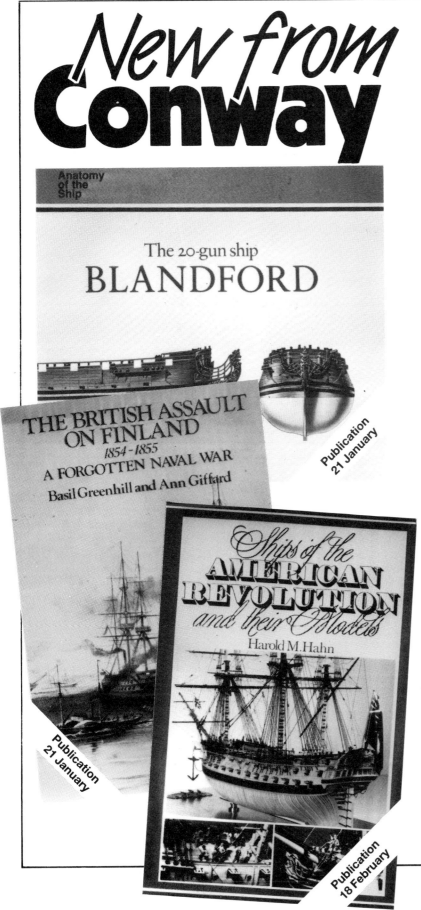

New from Conway

New 1988 Titles

ANATOMY OF THE SHIP THE 20-GUN SHIP BLANDFORD
Peter Goodwin
Blandford was the first in a long line of 20-gun Sixth Rate ships built for the Navy from 1719 onwards. Their significant, but little known role in the development of the frigate is outlined, along with all the technical details of the ship that readers of the 'Anatomy' series have come to expect.
240 x 254mm, landscape, 120pp, 20 photos, 300 line drawings.
ISBN 0 85177 469 5. **£14.00.**

THE BRITISH ASSAULT ON FINLAND 1854-55:
A Forgotten Naval War
Basil Greenhill and Ann Giffard
The Crimean War began and ended in the Baltic. This is the full story of the campaign there in which for the first time a steam battlefleet went to war. As well as surveying the naval and military actions of the campaign, the authors delve into the political, social and economic history of the area.
234 x 156mm, 336 pages, 32 illustrations. ISBN 0 85177 470 9.
£20.00.

SHIPS OF THE AMERICAN REVOLUTION AND THEIR MODELS
Harold M Hahn
Model builders have until now had a very limited choice of subject from this important period in American history. In this book Harold Hahn describes seven ships, both British and American, and deals with their history as well as giving step-by-step descriptions of the construction of each model.
270 x 206mm, 288 pages, 200 illustrations. ISBN 0 85177 467 9.
£20.00.

AVAILABLE FROM ALL GOOD BOOKSELLERS or by post from CONWAY MARITIME PRESS LTD, PO Box 10, Teignmouth, Devon TQ14 9HH (please add 10% for postage and packing for orders of £15 and over and 15% for those under £15).
Credit card sales (ACCESS/VISA) phone: 01-583 2412

Send for a free illustrated catalogue of Conway titles

Putnam Aeronautical

'The name Putnam on a work of aviation history puts it in a class apart...' *Lloyd's List*

Putnam publish a wide range of aviation reference works. Not only are the most sought-after titles being brought back into print but new titles are being added to the series.

Japanese Aircraft of the Pacific War
R J Francillon
Describes the most significant types of aircraft operated by the Army and Navy immediately before and during the Pacific War.
216 x 138mm, 586 pages, 435 photos, 95 GA drawings.
ISBN 0 85177 801 1. **£18.00**

Supermarine Aircraft since 1914
C F Andrews & E B Morgan
Traces the entire development of Supermarine from its earliest days with flying boats and amphibians to its postwar role in the development of jet aircraft for the Royal Air Force and the Royal Navy.
216 x 138mm, 416 pages, 300 photos, 78 GA drawings.
ISBN 0 85177 800 3. **£18.00**

United States Navy Aircraft since 1911
G Swanborough & P M Bowers
'Details every aircraft used by the US Navy and also the Coast Guard and the Marines... there is a vast amount of information...' *Lloyd's List*
216 x 138mm, 556 pages, 485 photos, 136 GA drawings.
ISBN 0 370 110054 9. **£16.50**

British Naval Aircraft since 1912
O Thetford
The fifth edition of this well-established work includes technical data, photographs and line drawings of every aircraft used by the Royal Navy from 1912 and includes descriptions of the Sea Harriers and Sea Kings which played a notable part in the Falkland's conflict.
216 x 138mm, 492 pages, 400 photos, 118 GA drawings.
ISBN 0 370 30480 2. **£15.00**

Curtiss Aircraft 1907-1947
P M Bowers
A complete history of the pioneer work of Glenn Curtiss and the subsequent companies which bore his name. Amongst other projects Glenn Curtiss built the American flying boat which led to the Royal Naval Air Service's family of Felixstowe patrol flying boats and the flying boat NC-4 which made the first ever aerial crossing of the North Atlantic.
216 x 138mm, 644 pages, 567 photos, 77 GA drawings.
ISBN 0 85177 811 9. **£20.00**

Lockheed Aircraft since 1913
R J Francillon
Recently updated and revised, this edition has a new section on Lockheed missiles up to the latest F-19 Stealth fighter as well as details of all the aircraft built by the companies from the Loughead Model G floatplane to present high-performance aircraft.
216 x 138mm, 576 pages, 430 photos, 72 GA drawings.
ISBN 0 85177 805 4. **£24.00**

Publication 21 April
Saunders and Saro Aircraft since 1917
P London
Traces the entire development of Saunders and Saunders-Roe aviation activity, including the seaplane floats and flying boat hulls built before the First World War, and the Walrus and Sea Otter Amphibians built during the 1939-45 war. The company's early origins and contributions to the design and construction of boats and marine aircraft hulls are also examined.
216 x 138mm, 384 pages, 253 photos, 21 GA drawings.
ISBN 0 85177 814 3. **£20.00**

Publication 19 May
Vickers Aircraft since 1908
C F Andrews & E B Morgan
Covers the aircraft designed and built by Vickers since the great shipbuilding and armaments firm was asked by the Admiralty to construct a naval rigid airship. Of special interest are descriptions of the designs of the postwar period, including the Viking amphibians and Viastra transports.
216 x 138mm, 576 pages, 526 photos, 99 GA drawings.
ISBN 0 85177 815 1. **£20.00**

Available from your local bookseller or direct from Putnam Aeronautical Books, PO Box 10, Teignmouth, Devon TQ14 9HH. (Please add 10% for postage and packing.) Credit card (Access/Visa) sales: 01-583 2412.

Send for a free catalogue of Putnam books

Don't Miss Our Superb Books

New
Cruisers of the Royal and Commonwealth Navies
A book long sought after .. now we can supply it! 256 pages. Hardback, well illustrated. Don't miss it! **£19.95**

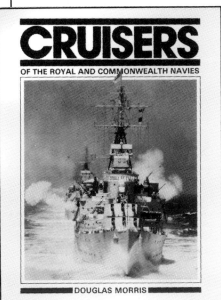

The Royal Navy at Portland Since 1845
A fascinating read, well illustrated with some very rare photographs. **£10.95**

HMS Hermes 1959-84
A superb giant commissioning book covering the entire career of the happy *Hermes*. Excellent photographs. One for the grandchildren? **£9.95** Hardback. **£5.95** Paperback

HMS Bulwark 1948-84
A similar book packed with nostalgia. If you served onboard you'll need a copy . . . well written by a former ship's officer. The ship has gone but the memory stays on . . . **£9.95** Hardback (only)

British Warships and Auxiliaries 1988
A new edition of our ever popular guide to the fleet (and F A A) just out. 112 well illustrated pages (inc colour). **£4.50**

The Royal Navy at Portland Since 1845
A new hardback — a great read for "Old Portland hands" — illustrated with some magnificent photographs. **£10.95**

Badgers and Battlehonours of H M Ships
Our "Classic" — all the badges in this superb hardback are in full colour. **£27.50**

We mail books worldwide . . . our prices include UK/BFPO post. Overseas add 10%.
Ask for our full FREE catalogue . . . Phone for more details of books above or our new naval video tapes . . . See our bargain offers listed too . . .
Have your seen a copy of our magazine "Warship World" — well recommended. Send just £1.50 for a current sample copy.

Maritime Books, Lodge Hill, Liskeard PL14 4EL
Tel: 0579 43663 (until 9pm)

Subscription Order

Take out a subscription to Warship
It's easier to subscribe. One year's issues for £14.00. There's no extra charge for postage and packing.

Subscription rates, prepaid. United Kingdom £14.00. Canada $35.00. USA $26.00. Other countries £16.00. These rates include surface post and packing. Airmail rates on request.

Payment
Visa or Access. Alternatively, cheques or postal orders, international money orders (Giro) or cheques in major currencies (overseas). All should be crossed and made payable to Conway Maritime Press Ltd, and endorsed 'A/C payee'.

Post
Post your order to: Conway Maritime Press, 24 Bride Lane, Fleet Street, London EC4Y 8DR. Alternatively telephone our credit card line: 01-583 2412.

Please send me four issues of Warship commencing with:

☐ No. 45 (January '88)   ☐ No. 47 (July)
☐ No. 46 (April)   ☐ No. 48 (October)

Name

Address

Please charge my credit card: £

Signature

Date

Please tick ☐ VISA   ☐ Access

card no:

WARSHIP

SMALL ADVERTISEMENTS

Press Day:- *Warship* is published on the first day of January, April, July, and October. Copy deadline is eight weeks before publication date.

Rates:- (Inclusive of VAT) Lineage, 30p per word with a minimum of 15 words, ie, £4.50. First 3-4 words set bold at no extra charge. Semi display, with box rule or border £8.00 per single column centimetre (minimum 2.5 cms £20.00). Display rates on application. Series discount 10% for four consecutive insertions.

Box Numbers:- Available for private advertisements only. £2.00. Please address replies to Box No., *Warship*, Conway Journals, 24 Bride Lane, Fleet Street, London, EC4Y 8DR.

Payment:- Prepayment by cheque or postal order made payable to Conway Maritime Press Ltd. and sent together with the advertisement copy to: The Advertisement Manager, Conway Journals, 24 Bride Lane, Fleet Street, London EC4Y 8DR.

Please Note:- The Publishers retain the right to refuse or withdraw advertisements at their discretion and do not accept liability for delay in publication or for clerical or printers' errors although every care is taken to avoid mistakes. It is the advertiser's responsibility to check that the first insertion in any series is published correctly and any corrections must be notified in time for the second insertion failing which the publishers will not accept liability or offer any refund of charges.

ANTHONY J. SIMMONDS
15 THE MARKET, GREENWICH
LONDON SE10 9HZ, ENGLAND
01-853 1727

We specialise in books on all aspects of the sea, naval and maritime history, naval architecture and art. Catalogues sent on request or visit our shop at above new address. Closed on Sundays.

THOUSANDS of military, naval, aviation and railway books in stock for world-wide mail-order; renowned service and packing. Free illustrated catalogue – call Midland Counties Publications on 0455-47091.

IMPERIAL WAR MUSEUM
WELCOME ABOARD HMS BELFAST
the friendliest museum afloat!

Explore all seven decks of this majestic old cruiser and discover what life at sea was really like during World War II.
Open 7 days a week from 11am. Admission charge. Underground: Tower Hill, London Bridge. Symons Wharf, Vine Lane, Tooley Street, London SE1 2JH. Tel: 01-407 6434.

ROYAL NAVAL MUSEUM HM Naval Base, Portsmouth, Hampshire PO1 3LR. Tel: 0705 822351.
The Entire History of the Royal Navy from Tudor Times to the South Atlantic Campaign of 1982. Special Display on the Ships and Operations of the Modern Navy.

ROBERT AND SUSAN PYKE Second hand and antiquarian maritime booksellers, large wide-ranging stock. Free catalogues. 2 Beaufort Villas, Claremont Road, Bath. Tel: 0225 311710.

NATIONAL MARITIME MUSEUM The History of the Sea in a Palace by the River. Romney Road, Greenwich, London SE10. Tel: 01 858 4422.

SEAFARERS: For Seagoing books – old and new. For a treat, why not visit our maritime shop at 18 Riverside Market, Haverfordwest, Pembrokeshire. The only seabook specialist west of the border – or phone for our list: 0437 68359.

OVER 5000 NAVAL & MARITIME BOOKS old and new, postcards, cigarette cards and other items of marine ephemera for sale. Catalogues sent on request. SAE to Michael Prior, 34 Fen End Lane, SPALDING, Lincolnshire PE126AD (tel: 0775 6185).

FISHER NAUTICAL have the largest, continuously restocked selection of books on the sea and ships. We also need to buy all types of nautical books and ephemera. Send for list. Fisher Nautical, Huntswood House, St Helena Lane, Streat, Nr Hassocks, Sussex BN6 8SD. Tel: 0273 890 820.

FLEET AIR ARM MUSEUM
ROYAL NAVAL AIR STATION,
YEOVILTON, SOMERSET
Tel. 0985 840565

See the exciting story of the Fleet Air Arm since the Royal Navy first took to the air in 1908 through to the Falklands War.

Open 10-5.30 (March to October)
10-4.30 (November to February)

ANTIQUARIAN AND SECONDHAND
BOOKS
FOR SALE BY CATALOGUE
NAVAL, MARITIME, SHIPPING,
YACHTING, etc.

ABBEY BOOKS
'HIGHVIEW', STEYNE ROAD,
SEAVIEW, ISLE OF WIGHT,
PO34 5BH
Tel. 0983 612821

MCLAREN BOOKS Scotland's leading dealers in out-of-print and scarce NAUTICAL BOOKS Antiquarian and Modern. Catalogues issued. Books purchased. 91 West Clyde Street, Helensburgh, Dunbartonshire G84 8BB. Tel: 0436 6453 or 820487.

MARITIME MODELS We are the leading specialist shop for the marine modeller and main stockist for Sirmar 1:96 modern warship hulls and fittings. List £1.75 inc p & p. Also warship plans. MARITIME MODELS, 7 Nelson Road, Greenwich, London SE10. Tel: 01-858 5661.

THE MARY ROSE SHIP HALL AND EXHIBITION See Henry VIII's warship in her new dry dock and the new exhibition featuring fascinating displays of the historical 'treasures' of the ship – the everyday objects she carried on a July day in 1545.
Open every day from 10:30-5:00. HM Naval Base, Portsmouth PO1 3LX. Tel: 0705 750521/839766.

JULIANS FOR NAUTICAL BOOKS. Send for our catalogue of secondhand books covering all aspects of the sea and ships: RN, MN and yachts. 35 Venn Grove, Hartley, PLYMOUTH PL35PH. Tel: 0752 701740.

CHATHAM HISTORIC DOCKYARD
MEDWAY

Visit Kent's Georgian Royal Dockyard and enjoy Britain's largest concentration of scheduled ancient monuments. See its unique traditional ropery in action.
April – October open Wednesday – Sunday 10 – 6. Admission charge.

MOTOR BOOKS NAVAL section stocks CONWAY's publications together with thousands of others. We also specialise in MOTORING, RAILWAYS, AVIATION & MILITARY. MOTOR BOOKS, 33 St Martins Court, London WC2N 4AL. Tel: 01 836 6728/5376.